DISASTER MANAGEMENT & Continuity of Operations Planning (COOP) HANDBOOK

From a Corporate, Federal and Local Government, Emergency Coordinator's Perspective

KENNETH L. BRISCOE
MA, Security Management & CBCP

Copyright © 2007 by Kenneth L. Briscoe

All rights reserved. No part of this book shall be reproduced or transmitted in any form or by any means, electronic, mechanical, magnetic, photographic including photocopying, recording or by any information storage and retrieval system, without prior written permission of the publisher. No patent liability is assumed with respect to the use of the information contained herein. Although every precaution has been taken in the preparation of this book, the publisher and author assume no responsibility for errors or omissions. Neither is any liability assumed for damages resulting from the use of the information contained herein.

ISBN 0-7414-3691-4

For briefing, presentations or program development support, contact:

Kenny Briscoe

E-mail: Kbri0606@aol.com

301-807-6060

Published by:

INFINITY
PUBLISHING.COM
1094 New Dehaven Street, Suite 100
West Conshohocken, PA 19428-2713
Info@buybooksontheweb.com
www.buybooksontheweb.com
Toll-free (877) BUY BOOK
Local Phone (610) 941-9999
Fax (610) 941-9959

Printed in the United States of America

Printed on Recycled Paper

Published April 2007

Contents

The Age of Emergency Management 1
Emergency Operations Center (EOC) & Alternate Site Concepts 16
The FEMA (FPC 65) 33
Business Continuity Management 53
Security & Business Continuity 67

Appendixes
A. Emergency Work-force letter 78
B. Occupant Emergency Action Plan 81
C. Example of Receiving and Reporting Bomb Threats 89
D. Example of a Sheltering in Place 95
E. A Check List for Your Business Recovery Plan 100
F. Example of a Business Continuity Plan 104
G. Example of an Alternate Site Guide 112
H. Example of Alternate Site Activation, Reception and Notification Checklists 134
I. Example of FEMA, Continuity of Operations Plan Template Instructions 141
J. Example of a Mission Essential Function (MEF) Analysis 153
K. Example of a Business Impact Analysis (BIA) 154

Chapter 1

The Age of Emergency Management

The Universal Prospective

Do you recall watching the news in the aftermath of a disaster, which depicted armed soldier everywhere while hoards of people tried to acquire supplies such as water, ice and charcoal just to support their daily needs? In the back of your mind, you probably said to yourself, "What if something like that happened here. I need to get some emergency supplies". Well it's no longer just a TV news broadcast that you're` watching. In recent times, these types of disasters have become very real to many and maybe you have been there.

Disasters are forces of nature that just happen. But we can't live our lives worrying about things over which we have no control. In fact, in past times, the common philosophy was to 'just deal with it'. There were no plans, no advance forecast and no preparation for surviving a disaster. The response to a disaster was usually reaction, as opposed to proactive. Yet, the aftermath of a disaster resulted in the survivability of those affected by mean of scrounging in order to acquire whatever was needed, (i.e., ice, plastic, construction supplies, etc) at any cost. New businesses were formed overnight and prices skyrocketed in order for vulturous individuals to profit from the misfortune and lack of planning for disasters. Not to mention insurance. Were you covered? Did you read the fine print, which detailed what was NOT covered? Two words...Hurricane Katrina!

During previous disasters, anarchy resulted, which lead to the U.S. Army or the National Guard being called in to restore order and control. The U.S. military missions supporting national disasters, along with the availability of manpower and assets appeared to be the ideal fit to manage large –scale emergency operation. *However, our military is being pushed to the limit and needs our help!*

Disaster Statistics

According to statistics published by the Association of Records Managers and Administrators, 44 percent of businesses directly affected by a disaster never re-opened for business after losing their records during the disaster. While 30 percent of businesses did reopen, the business failed to survive beyond three years after the disaster. Data on floods and hurricanes are even worse. Failure to plan statistically cuts the business chances of survival in half.

Trying to operate a business after a disaster is difficult, to say the least. Restoring the daily functionality of a business is time consuming and requires planning. Coping with daily management issues, employees issue and let's not forget insurance companies,

all the while trying to recover from the disaster, can easily become more than most can tolerate.

In 2005, although the death toll from Hurricane Katrina was less than 1,000, the loss of life combined with the loss of property makes this one of the worst natural disasters in American history. Less than a year after Katrina, the Great Tsunami that killed about 200,000 in Asia. This unfortunate series of events has many wondering whether the world is facing much more unpredictable and intense weather, possibly due to global warming. In every disaster, communities and businesses are impacted near and far, directly or indirectly.

Plan for the Worst

When you think about risk management and disaster preparedness, developing a disaster recovery plan for a business should include frameworks for the most likely disasters that may affect the area. In other words, a business wouldn't likely plan for recovery from a hurricane or tornado when the business doesn't exist in a region where the occurrence of these force of nature is infrequent or non-existent. However, be flexible.

Emergency Management / Business Continuity & Emergency Planners

Today's emergency planners are the backbone of an organization. They effectively prepare the company and its' clients to survive disasters whether manmade or natural and enable them to continue daily business operation. Emergency planner may be the key to your business survival. Unfortunately, many senior mangers may not agree with this statement, not fully understanding that the failure to plan for a disaster today may result in costly expenses tomorrow.

Companies are downsizing across the board. Therefore, understanding the company's Mission Essential Functions (MEF) and the Business Impact Analysis (BIA) are required elements to succeed in today's unstable business environment. MEF's are the critical functions, without which, the business cannot survive. They consist of those functions that must be performed without unacceptable interruption to achieve the company's critical missions. They include organizational command and control of assets, performing other operations that must be carried out to achieve mission success, receiving, assessing, analyzing, processing, displaying, and disseminating information necessary to perform critical missions and support decision making, and processing vital information for customers, contractions, and employees.

BIA is defined as analysis that identifies the impacts of losing organizational resources. The BIA measures the effect of resource loss and rising losses over time in order to provide an organization with reliable data upon which to make decisions on risk mitigation and continuity planning.

    (Appendix K)

Figure 1

Management must know its' organization posture in-order to develop a strategy to mitigate possible disruption of critical business functions.

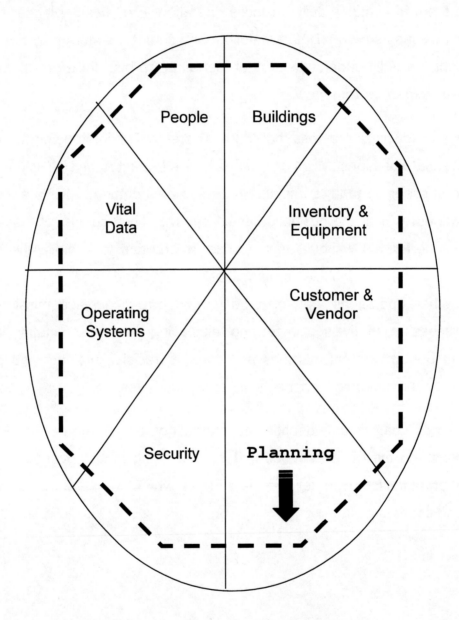

(Appendix J is a MEF Function Assessment Form that reviews and assesses your organizations Mission Essential Functions (MEF) in support of your Business Continuity Plan. Each element will use the most recent operational activities to review their organization's functions.

Business Risk Management

Risk is defined as 'the perceived extent of possible loss'. Businesses will have different views of the impact of a particular risk in their operations. However, your organizational risks should be derived from the BIA and its Recovery Time Objective (RTO). Trivial risk for one organization may destroy the livelihood of another. *Planners need to define the organizational risk.* Protecting information and defending a computing infrastructure is a great place to start.

Remember that the organizations security plans are an integral part of the risk assessment. The assessment must balance the relative value of the organization's assets against the cost of protecting them and against the probability that they will be violated. The first step of business impact analysis is to evaluate the effects of disasters and security risk on your business, identifying the areas of greatest vulnerability. Afterward, develop a risk management process driven by your impact analysis, giving you the most bangs for the buck.

Figure 2

Risk Cause and Effect

Threat → **Vulnerability** → **Risk**

>>>> The Storm >>>>>>> Are you Ready >>>>>>> Your Assets >>>>>

Risk Analysis, How to evaluate and manage the Risks is our Business.

Business Risk Analysis is a plan that helps to assess the risks that an organization may confront in daily operations and/or during a disaster. A good risk analysis will help your organization decide what actions to take to minimize disruptions of your operations. It will also help you to decide whether the strategies you could use to control risk are cost-effective. Senior Management must understand possible risks and strategies to resolve disaster issues. Having a Business Recovery Plan to mitigate these problems is management's responsibility.

Conduct a Business Impact Analysis (BIA) – the BIA is a key step in the continuity planning process. *See Appendix K*

Today's Emergency Planning Coordinators (EPC) have two key focal points: Disaster Recovery and Business Continuity. In the past, Disaster Recovery was the basic focus of emergency management. Fitting structures, getting power and basic life needs were the key issues for a large percentage of businesses and households. However, today's recovery personnel are primarily concerned with system recovery. Information technology has not only changed our day-to-day lives, but has changed emergency management recovery priorities; this was inevitable.

A Business Continuity Plan includes Information Technology (IT) disaster recovery planning in other words, the total business perspective. Continuity plans are built upon vulnerability and risk assessments developed by disaster planners. Continuity planning is a term for planning the continuity of an organizations operational and business functions, should disruptions occur due to natural disasters and manmade causes or technological failures. This all-purpose planning focus is a necessary action to ensure the performance of essential business functions. This includes business and operational continuity planning, IT disaster recovery planning, emergency and disaster preparedness planning, and any other planning related to maintaining the capability to conduct organizational functions before, during, and after disruptions.

Business leaders should take actions to ensure the continued operation of information networks and systems vital to the continuity, survival, recovery, and reconstitution of their MEF without unacceptable interruption. Business Continuity planners should identify and prioritize MEFs and identify those that can be deferred during a major emergency or other emergency affecting the continuous performance of those functions until time and resources permit restoration. The organization should consider the impact on the business, organizations, customers and employees when identifying functions that can be deferred. There are certain questions the management must address:

What needs to be protected?

What do they need to be protected from?

What are you already doing to protect them and what more could you do?

What rules or procedures are currently in effect to prevent or minimize the impact of a disaster?

The bottom line is every business needs a disaster planning process in place that includes three types of plans: Mitigation, Recovery and Business Continuity Plan.

Mitigation is defined as having activities to limit or eliminate the possibility of loss of life and/or property. Recovery is defined as activities to salvage and re-establish operations. Business Continuity Plan is defined as continuing to do business during adverse conditions.

How long should the recovery period be?

How long can the business remain closed with limited access to its vital information or continue operations without computers?

Would a few days of second-rate service be acceptable to the customers who depend on your business?

The answer to the question regarding how long the recovery period can be has obvious implications for how much emphasis should be placed on mitigation and continuity planning. If you cannot afford to be out of business, you need a workable plan to ensure that your company can continue to function after a disaster occurs. Many businesses assume that their customers will still be there after the period of recovery is over, even if recovery is a long and slow process. This theory is idealistic for most perspective customers.

Understanding COOP and Emergency Management Terms

When discussing emergency management, there are many different terms and definitions addressing the same issues. Terms recognized by federal, local, corporate and industrial organizations are frequently different. Understanding terms and working together is a key factor in fighting emergency issues and protecting this great nation.

As government and corporate America address emergency issues, we must be on the same page. For example, in the Nation's Capitol, which organization makes the final discussion in the National Capitol area in regards to emergency management? Do you know?

Continuity of Operation (COOP) is normally a federal government term, and Business Continuity is normally a corporate term. However, the functions are the same. Executing recovery actions for federal and corporate organizations includes; (1) what is needed to recover, resume, continue or restore business functions; (2) where to go to resume business and operational function; (3) when business functions and operations must resume and having detailed procedures for recovery, resumption, continuity, and restoration.

Figure 3

Understanding the Federal Government Continuity Program:

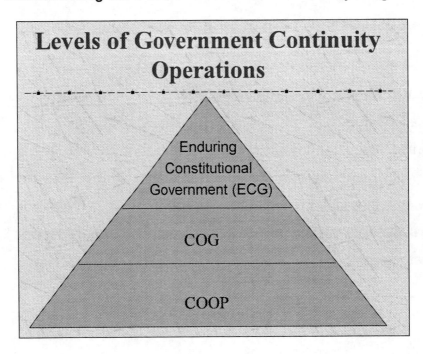

Definitions:

The highest level of the Federal Government Continuity Program is "Enduring Constitutional Government (ECG)." ECG is a cooperative effort among the Executive, Legislative, and Judicial Branches of government, coordinated by the President, to preserve the capability to execute constitutional responsibilities in a catastrophic crisis. ECG is the overarching goal; its objective is the preservation of the constitutional framework under which the nation is governed. ECG requires orderly succession, appropriate transition of leadership, and the performance of essential functions by all three branches of government. ECG is dependent on the effective branches, *Continuity* of Government (COG) and COOP, and capabilities.

Continuity of Operations (COOP) is an internal effort within individual components of the Executive, Legislative, and Judicial branches of government assuring the capability exists to continue uninterrupted essential component functions across a wide range of potential emergencies. These include localized acts of nature, accidents and technological or attack-related emergencies. COOP involves plans and capabilities covering the same functional objectives of COG and must be maintained at a high level of readiness, and be capable of implementation both with or without warning. In this sense, COOP is not only an integral part of COG and ECG, but is more simply "good business practice" - part of the Department of Defense's fundamental mission as a responsible and reliable public institution.

Continuity of Government (COG) is a coordinated effort within each branch of government ensuring the capability to continue its minimum essential responsibilities in a catastrophic crisis. COG is dependent on effective continuity of operations plans and capabilities. COG activities involve ensuring the continuity of Mission Essential Functions (MEF) through plans and procedures governing succession to office; emergency delegations of authority (where permissible, and in accordance with applicable law); the safekeeping of vital resources, facilities and records; the improvisation or emergency acquisition of vital resources necessary for the performance of MEF; and the capability to relocate essential personnel and functions to an alternate working site(s) and sustain performance of MEF at alternate work site(s) until normal operations can be resumed.

Emergency Terms:

Incident Command Post is defined as the location or locations where the on-site incident command functions are performed. *Incident Management Team* is defined as the Incident Commander and appropriate staff assigned to an incident on site and at the EOC. *Emergency Operations Center* is defined as the location where emergency response and recovery activities (supporting the Incident Commander) and selected organizational operations are conducted.

Emergency Planning Coordinator (EPC) is defined as a person assigned to develop, exercise and update the emergency plan at a particular work site.

IT Disaster Recovery Planning is defined as a process that focuses on the data/computing center and/or local and wide area network recovery following a disruption including specific actions for restoring or recovering IT and other systems after they fail. These plans usually include the procedures to safeguard information by conducting backups or similar procedures to permit the restoration of information. These plans are prepared by system administrators, but include appropriate links to the business continuity plans of all functions that rely on that system or IT component. (See also Information Assurance.)

Emergency Planning is defined as procedures for fires, bomb threats, hazardous material releases, etc., which deal with how emergency response is conducted at a particular work site.

Disaster Planning is defined as the procedures for storms, earthquakes, and other natural disasters. Most often used in conducting mitigating actions, such as evacuations, when warning conditions exist. Continuity planning is defined as planning done before a disaster to ensure that the business can resume functions as soon as possible after the disaster.

Incident Response Plan. The Incident Response Plan establishes procedures to address cyber attacks against IT system(s). These procedures are designed to enable security personnel to identify, mitigate, and recover from malicious computer incidents, such as unauthorized access to a system or data, denial of service, or unauthorized changes to system software (e.g., malicious logic such as a virus, worm, or trojan horse).

A Mitigation Plan is defined as a plan to minimize damage and prevent an emergency from becoming a disaster by taking specific actions before an emergency occurs. Mitigation is action taken before a disaster to reduce the risk of disaster or reduce the impact of the event. A mitigation plan is a list of mitigation activities with priorities, commitments, and a timeline. Once a mitigation plan is written, it should be reviewed regularly to evaluate how well mitigation goals are being met.

Requiring more time and money does not make an action impossible. It simply means that you also have to plan how to get the money. Increasingly, agencies such as the Federal Emergency Management Agency (FEMA) are promoting mitigation because they have done the math. They know that mitigation saves money in the long run. They also know that mitigation can save lives.

Mitigation activities cannot be done to help your business until there is a plan that details what should be done as mitigation. The Mitigation Plan is part of the Business Continuity Plan. *Business Continuity Plan* is the primary plan for a business emergency. However, the BCP includes the following plans: The Mitigation Plan, the Response Plan, and the Recovery Plan.

Disaster response plans provide for immediate reaction after a disaster occurs. Response plans are directed toward the safety of people and company assets. Plans are required for evacuating the building and coping with other life and safety emergencies as required. Whether it is a fire, an explosion, a chemical spill, a tornado, flooding, or some other disaster, response plans are imperative for management, employees and customers to address sound and life-threatening decision during dangerous times. The safety of people has to be the top priority.

Disaster response plans, whether they focus on life-threatening issues or company assets, must be written as clear, "easy to use" instructions for the first response team to use as they begin to cope with an emergency.

Government Emergency Terms:

Emergency Relocation Group (ERG) is defined as, staff identified to relocate to secure facilities, maintain command and control, execute MEFs, reorganize and redirect resources, conduct required interagency coordination and implement decisions.

National Security Emergency (NSE) is defined as any occurrence including, but not limited to, natural disaster, military attack, technological failures, civil unrest, or other disruptive conditions that seriously degrades or threatens the national security of the United States.

Operations Management Team is defined as the organization that coordinates management of support to customers in a COOP environment, orchestrating the efforts of the MEF cells and mission status reporting.

Continuity Response (Recovery) Cell (CRC) is defined as staff members assigned to manage and support execution of an agencies continuity functions, including relocation, operations, alternate facility support and follow-on continuity planning.

Crisis. defined as an incident or situation involving a threat to the United States, its territories, citizens, military forces, possessions, or vital interests that develops rapidly and creates a condition of such diplomatic, economic, political, or military importance that commitment of U.S. military forces and resources is contemplated to achieve national objectives.

Critical Infrastructure Protection (CIP) is defined as the identification, assessment, and security of physical and cyber systems and assets so vital to the nation that their incapacity or destruction would have a debilitating impact on national security, national economic security, and national public health and safety. Within the Department of Defense, it is the identification, assessment, and security enhancement of physical and cyber assets and associated infrastructures essential to the execution of the National Military Strategy. CIP is a complimentary program linking the mission assurance aspects of the Anti-Terrorism, Force Protection, Information Assurance, Continuity of Operations, and Readiness programs.

Information Assurance (IA) is defined as information operations that protect and defend information and information systems by ensuring their availability, integrity, authentication, confidentiality and non-repudiation. This includes providing for restoration of information systems by incorporating protection, detection, and reaction capabilities.

Mission Essential Functions (MEF) are the specified or implied tasks, which are required to be performed by statute or executive order, and are those continuing activities that must be performed without interruption to achieve critical component missions.

National Capital Region (NCR) is defined as the geographic area located within the boundaries of the District of Columbia; Montgomery and Prince Georges Counties in the State of Maryland; Arlington, Fairfax, Loudoun, and Prince William Counties and the Cities of Alexandria, Fairfax, Falls Church, Manassas, and Manassas Park in the Commonwealth of Virginia; and all cities and other units of government within the geographic areas of such districts, counties, and cities.

Disaster Inventory Storage

You must decide where and how to store the supplies. Furthermore, you must consider the vulnerabilities in and around your facilities. Avoid open outside areas if possible. Examine internal areas and avoid areas that leak, mold or have poor ventilation. Your storage area should have an alternate means of lighting, preferable skylights or windows. Internal closets are often good locations to secure equipment. Placing supplies on wooden pallets is good for ventilation and prevention of water/flood damage. Sealing supplies in shrink-wrap with an inventory list will also protect inventory from water damages. Keep the containers in one location. One of the most reliable means of storage is a small metal crate (10' by 10' or 20' by 20'). The container can be pre-loaded and is great for securing equipment or supplies. Thirty to 55 gallon metal or plastic cans can also be used for storage and easy transport to an emergency location.

Remember, your organization needs supplies and equipment for immediate response to a particular emergency. A few supplies to respond to water leaks would include; squeegees, plastic sheeting, mops, buckets and sponges, a wet/dry vacuum, portable generator, work lights, electrical cords, small tools and personal protective equipment such as plastic groves, plastic shoe covers, etc.

If you have a large stockpile, you may want to pack the containers by category, putting tools in one category, food and water in another, mold treatment supplies, drying supplies in another, and so forth. During disasters, many organizations purchase footlockers, which are great for packing items for expedient movement. Disaster supplies may need immediate access. Key Emergency personnel should have a key in their possession, or store a set of the keys with each BCP. The footlockers or storage containers should be labeled with an inventory list any items are removed for use they should be replaced. Furthermore, the BCP should have instructions on how to get access to the storage locations. Be sure all members of the disaster team or other emergency personnel know how to respond to an emergency, are aware of the locations of the kits and know who to contact and how to access supplies.

Off-site Storage

Off-site storage is a great option. However, if you do so, ensure that the storage is easily accessible and is available 24 hours a day, 7 days a week. Consider the driving distance, the location and whether the off-site location is in a dangerous area. Is the structure of the facility housing your supply sounded and sheltered? It is generally good to keep the most critical supplies in your own disaster kit at the location where the supplies are needed.

If you decide to create a large stockpile to handle major disasters, you may store some of the more cumbersome equipment (generators, dehumidifiers, and so on) outside your building. In addition, you might keep a limited quantity of supplies in your building and back-up supplies elsewhere. Some public institutions use a city or county warehouse for this purpose. However, be sure you know how to get access to these facilities.

Inventory

Inventory the supplies both on and off-site on a regular basis to determine that all materials are present and in good condition. This should be done at least quarterly, but monthly is preferable. If possible, seal and secure the supplies; you will be able to tell at a glance whether anyone has used them. Shrink-wrap also provides quick indication that the inventory has been opened. Furthermore, if the inventory is maintained in a closed and locked container, the likely hood of pilferage is less probable. Keep in mind that some materials need to be replaced periodically, such as batteries, tape and other limited shelf life items, so you need to establish a schedule for inspecting and replacing them.

Disaster Supplies

As your company develops your disaster supplies inventory, you must consider what disaster or disasters can occur in your area. Disaster supplies are versatile; however, focus on the most likely disaster for your area. Disaster supplies have a broad perspective from plywood to paper cups. During a disaster, plan on long hours, food, and sleeping considerations, but most importantly know your employees and respect their limitations.

Chapter 2
Emergency Operations Center (EOC) & Alternate Site Concepts

What is an EOC?

An Emergency Operations Center (EOC) is the central location for directing emergency operations during major emergencies and disasters. An EOC is equipped with the necessary communications equipment, emergency information and operational supplies to allow an organization to coordinate emergency response activities in order to ensure that all needs are addressed with the available resources. The management and staffing of the EOC is overseen by the Emergency Management Coordinator or its designated person.

A good example of an effective EOC facility consists of the following functional areas: an operations area, communications area and emergency management offices. The EOC building also houses other support areas, including a supply and copier room, kitchenette and restrooms. The EOC has a secured entrance and back-up power is provided by a generator. The EOC and communications room has heating and cooling systems independent from the rest of the facility. Communications systems include multiple telephone and dedicated fax lines, low band, VHF, UHF and satellite feeds of the National Weather Wire Service, CNN and weather radar; and a dedicated telephone circuit on the National Warning System (NAWAS).

The EOC is manned 24/7 and the daily functions are managed through a battle rhythm crisis management process, See Figure 4. Events and activities are monitored and tracked during the entire process.

Figure 4

This diagram depicts a common "Battle Rhythm/Crisis Management" process or a 24/7 manning of the EOC, which monitors all reports, events and critical information required to track improvements and incident management.

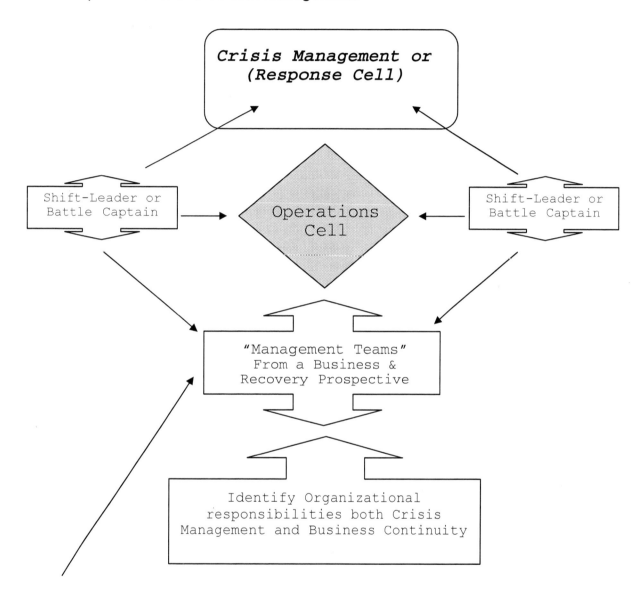

```
What are your organization's process
for the following:

How is information going to be
distributed & disseminated?
How is information going to come in?
What is the decision making process?
```

What are your MEF, and how will your organization continue to operate?

17

Figure 5

The concept and operating parameters for EOC operations during different threat level.

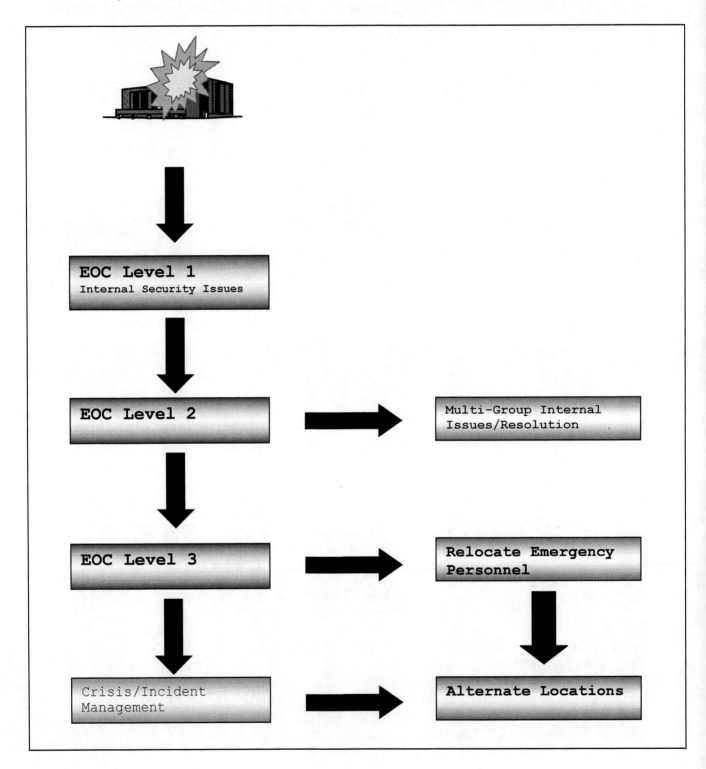

Three Levels of EOC Activation

Level 1: The EOC is activated within one of your organization's buildings and has limited staff consisting of Security personnel only. The event is such that no outside support or coordination is needed.

Level 2: The EOC is activated within your organization's buildings/facility and partially staffed with Security and other agency personnel. The event is such that possible protective measures are needed for your organization personnel by either in-place sheltering or site evacuation. The location will be in a your compound or facility.

Level 3: The EOC is fully activated and fully staffed by Battle Staff due to an event. During this level your buildings are inhabitable and your organization's Emergency Response Personnel must move to one of your alternate locations. See Figure 6

Relocate Emergency Personnel

Alternate Locations

Level One

Level One EOC will be activated at the discretion of the Director of Security and will be activated within one of your organization's buildings depending on the incident. Members will be notified by security recall roster. Emergencies such as severe weather alerts, fire alarms or power outages may require activation of Level One EOC at the discretion of the Director of Security.

Level One EOC Members

The EOC is activated within one of your organization's buildings and has limited staff consisting of security personnel only. The event is such that no outside support or coordination is needed. Level One EOC members include the Director of Security and his Department Heads, Operations, Team Leaders and Security Managers, as needed.

The members identified will be required to respond as required and will meet at a designated area to set up the EOC. Additionally, other members will be added as necessary to address a particular incident.

Level Two

The EOC will be activated at the discretion of the Director of Security and will be activated within one of your organization's buildings depending on incident. Members will be notified by recall roster. For incidents and accidents that will require Level Two activation, responding Emergency Response Official (s) will notify the Incident Commander until the scene can establish a Command Post (CP) and will assume responsibility is safely returned to your EOC control. Once a CP has been established, your organization liaison will post there and relay the status of the crisis back to the EOC to keep the command and employees apprised of the situations. The Incident Commander will have control of the resources at the scene and will have the ability to request additional resources from local responders. Scenarios such as structural fires, a shooting, hostage situation, a bombing, or HAZMAT release would require activation of Level Two.

Level Two EOC Members

The EOC is activated within your organization's facility and partially staffed with security and other agency personnel. The event is such that possible protective measures are needed for your organization personnel by either in-place sheltering or site evacuation.

Level Two EOC members include the Director of Security and his Department Heads, Operations, Team Leaders and Security Managers, as needed. In addition, depending on the incident, other EOC members may include The Safety Manager, Chief, Command Information, Human Resources Officer, Chief of Facilities, and other members as the situation dictates or requested by the Senior Management.

Level Three

The EOC will be activated at the discretion of the Senior Management. Level Three EOC will relocate your organization's personnel to an alternate site. For incidents and accidents that will require Level Three, responding Emergency Response Official (s) will notify the Incident Commander and establish a Command Post (CP) and will assume responsibility of the incident. Once a CP has been established, your organization liaison will post there and relay the status of the crisis back to the EOC to keep the command and employees apprised of the situations. The Incident Commander will have control of the resources at the scene and the ability to request additional resources from local responders.

Level Three EOC Members

Only Key or mission essential personnel are required to relocate. Many of your personnel will telecommute, work from home or work from one of your organization's alternate sites. The Safety Manager, Chief Information Systems, Human Resources Officer, Chief of Facilities, other members and Senior Management may be required as the situation dictates or require critical personnel.

Concept of Operations

The Business Continuity Planner has responsibility to develop plans for emergency management for your organization. During an incident, the Security Division will provide first response and incident command until requested outside resources arrive. Once outside responders arrive, the EOC will continue to operate to assist with the management of the crisis.

At the onset of an incident, the Director of Security or his designee will be the on scene command representative who has the sole authority to relay information to the EOC. In consultation with senior management, other EOC members, and outside authorities including security will have the authority to decide if the incident warrants activation of the EOC and where the EOC will be located. The purpose of the EOC is command, control, and communications. This is accomplished by gathering information from the incident scene, relaying information to senior management and designated officials, setting policy and making operational decisions, and disseminating information to your organization's staff. One member will keep a log of events to assist in reconstruction of the incident and to assist in identifying after action items as necessary. The EOC will be activated until the decision is made that the crisis warrants a Level Three activation of the EOC.

The Recovery Mission

Disaster Recovery Activities and programs are designed to return the entity to a customary operational condition. Before discussing Disaster Recovery, businesses must address the basic issue of Risk Management. Today's business environment is largely focused on Information Technology (IT). Disasters come in numerous forms to include manmade and natural: Internet failure, chemical and biological agents, workplace violence, fires, acts of terrorism, construction problems, hurricanes, tornadoes, and floods. Anything that risks human health and the environment, etc., are all forms of disasters.

Risk Analysis – Risk analysis is broadly defined to include risk assessment, risk communication and risk management. In the IT environment, risk is the process by which a company analyzes the business and technology necessary to continue to provide or return vital systems to standard operational levels. This process is important to business survivability. Risk must be analyzed concerning employees and their families and customers. Furthermore, the protection of the positive image of your company is important.

Planning Considerations For Disaster Recovery:

Considerations in risk assessment and the first stages of disaster planning include:

What needs to be protected?

What does it need to be protected from?

What are you already doing to protect them—and what more could you do?

How long can the recovery period be?

How much time and money can be spent to ensure that the business can function after a disaster?

Can the business afford a disaster?

Disaster Recovery Teams and Their Responsibilities

Emergency Response Commander & Staff (ERC) - directs all business recovery operations. The ERC is responsible for general supervision of re-establishment of the business infrastructure. The ERC is the officer in charge of the Emergency Operation Center, and reports significant events and updates to the Command Center (C2). The ERC has senior management decision-making authority. Therefore, the ERC is a key Operations Manager in all recovery activities. The ERC's Duties are many, a few are listed below.

- Develops and regularly updates disaster preparedness plans for protection and recovery efforts.
- Plans initial damage assessment and establishes priorities of recovery efforts.
- Activates the EOC and notifies Senior Management of emergency and Operations
- Manager of support needs.
- Issues daily situation report to Senior Management.

- Stores and issues equipment and disaster supplies from storage.
- Identifies and protects vital records.
- Thinks out of the box to take immediate action to reduce or eliminate risk of damage to business.
- Assesses the extent and possible types of damage that may occur to the business as a whole and/or as a subdivision.
- Refines recovery activities and priorities based on type and extent of damage and establishes a priority list for efforts based on BIA.
- Pre-appoints unit heads as needed to supervise any part of the recovery operation.
- Activates, supervises and when needed trains recovery crews.
- Determines the sequence and methods of recovery.
- Establishes work areas for all parts of the recovery operation, with assistance from the procurement officer, facilities manager and others necessary for the provision of space, supplies and equipment.
- Establishes safe storage locations on-site and off-site, as directed by BCP.
- Authorizes discard of damage equipment.
- Initiates plans for long-term clean up and restoration.

Site Damage Assessment Coordinator/Team (SDAC) - coordinates and assesses all damage activities and report findings to ERC. The SDAC duties are many, a few are listed below:

- Reports initial damage assessment to the ERC and receives initial directions.
- Takes immediate action to reduce or eliminate risk of damage to business and personnel.
- Obtains emergency supplies as necessary and advises ERC of additional supply and equipment needs.
- Coordinates with photographer to document the disaster.
- Advises ERC on the sequence and methods of salvage.
- Activates, supervises and as needed, trains damage assessment work crews.

- Gives specific direction to safety agencies and staff assigned to the recovery effort as required.
- Recommends on-site and off-site storage areas to ERC.
- Develops business resumption plans which are mechanisms for providing access as, soon as possible through means such as setting up off-site service points or re-opening a portion of the building or office.

Data Recovery Division (DRD)/Team - responsible for the protection and recovery of the business' vital information, mainframes, PC, etc. The DRD reports vital data and electronic records issues to the ERC. Furthermore, the ERC reports finds and issues to senior management. The DRD, as part of preparedness, oversees routine weekly/daily system backups, provides for off-site storage of backup copies, and identifies potential off-site facilities that could be used in the event of a disaster. Supervises or contracts for salvage or restoration of data processing equipment, software, and files. Plans and manages the relocation and maintains off-site locations and systems, hardware and software. The DRD is the ERC's main source of information technology for the reconstitution effort.

Alternate Site Development

Having an alternate working location is critical to business continuity. Alternate sites may make a difference in your organization's survival during a disaster. The primary objective of an alternate site is being able to conduct "normal business" without your customer feeling the burden. An alternate working location will further support a smooth transition to the reconstitution and recovery of an organizational process.

In determining an alternate site for your organization, consider the physical location: In the case of a local emergency or threats, the alternate site should be within a 60 mile radius from your location. In the case of a widespread emergency or threat the alternate site should be within a 60-150 mile radius from your primary location. The alternate site should be capable of supporting your operation 24x7 in a threat-free environment.

There are numerous possible alternate site concepts, the three most common being: Hot, Warm, and Cold Sites.

A Hot Site has all functionality to include systems, communication and life support. It is the most expensive to maintain. Advance support agreement and contracts are required. Furthermore, Hot Site support systems requirements are configured with necessary hardware, supporting infrastructure and support personnel and is typically staffed 7X24.

- Mobile Sites – these are self-contained, transportable shells custom-fitted with specific telecommunications and IT equipment to meet system requirements.

- Mirrored (Redundant) Sites – these are fully redundant facilities with full, real-time information mirroring your normal work stations and operations.

A Warm site, servers, and required infrastructure are in place. However, further hardware and operational support is required. Warm sites are partially equipped office spaces that contain some or all of the system hardware, software, telecommunications, and power sources. The warm site is maintained in an operational status ready to receive the relocated system. The site may need to be prepared before receiving the system and recovery personnel.

- Shared Sites are equipped office spaces that contain some or all of the system hardware, software, telecommunications, and power sources. However, a short time frame is required to activate this facility. These workstations are normally utilized by a non-essential person for daily operations. However, your personnel will take the place of the normal operator, and the normal operator will depart the area.

A Cold Site. An empty warehouse is a good example; you're starting from zero. Cold sites typically consist of a facility with adequate space and infrastructure (electric power, telecommunications connections, and environmental controls) to support the IT functions. It may contain a raised floor and other attributes, but it does not contain IT equipment and usually does not contain office equipment (phones, faxes, copiers).

All agencies shall designate alternate operating facilities as part of their BCP plans, and prepare their personnel for the possibility of unannounced relocation of essential functions and essential personnel to these facilities. Facilities may be identified from the existing agency, local infrastructures or external sources. Facilities should be capable of supporting operations in a low threat environment, as determined by the geographical location of the facility, a favorable assessment of the local threat, and the collective protective characteristics of the facility. In acquiring and equipping such facilities, agencies are encouraged to consider cooperative interagency agreements and to promote sharing of identified alternate facilities. However, care should be taken to ensure that shared facilities are not overcommitted during incidents. Alternate facilities should provide:

(1) Sufficient space and equipment to sustain the relocating agency
(2) Capability to perform essential functions within 12 hours of activation and for up to 30 days under various threat conditions, including threats involving weapons of mass destruction
(3) Reliable logistical support, services and infrastructure systems
(4) Consideration for the health, safety, security and emotional well being of relocated employees
(5) Interoperable communications with all identified essential internal and external organizations, customers and the public
(6) Computer equipment, software and other automated data processing necessary to carry out essential functions.

There are three critical plans required for alternate site development

Notification Plan for relocating and On-Site Personnel
Site Activation Plan, Setting-up your alternate Site
Site Reception Plan, Receiving relocating personnel

Figure 6

This is a "Possible Alternate Site Concept" for your organization.

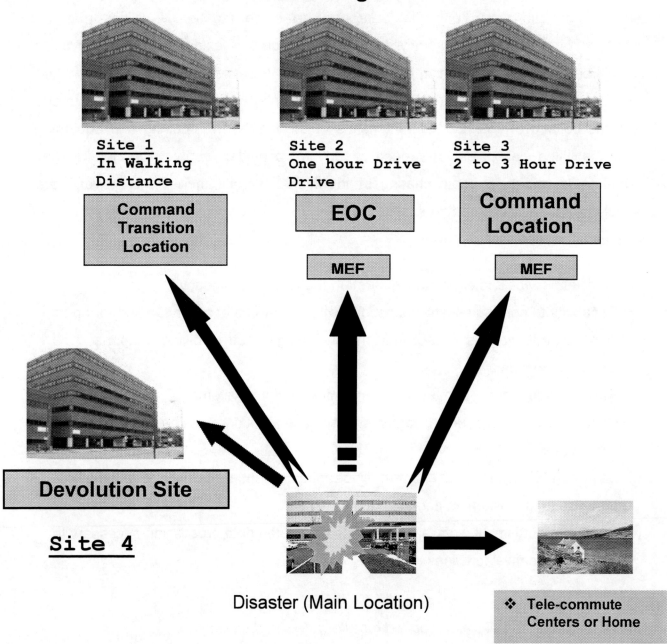

Site 1 depicts a location within walking distance from you main office. If your office building is inhabitable, this location is good for a transition point for your organization.

Site 2 depicts an alternate Site EOC, is a one hour drive; therefore, your personnel may commute from their homes.

Site 3 depicts an alternate Site, a two to three hour drive; total support is required. (See Appendix G. Alternate Site Guide)

Non-Essential Personnel may work from home or telecommute.

Devolution Site 4, a good example is Hurricane Katrina, a Government Devolution Site provide your last hope for command and control, "Use in the event that all else fails".

Disaster Location or the Incident Command Center (ICC) ensure effective and quick coordination of actions in response to disasters effecting your organization. The ICC personnel are positioned at the (Location of the Incident). ICC provides a comprehensive framework to coordinate actions when many agencies and organizations are involved in responding to an emergency. For ICC to work, there has to be a commander who has the authority over all the participating agencies and organizations, and this person is referred to as the Incident Commander, normally the local Fire or Police department on the seen.

See Appendix –G. Alternate Site Guide, for alternate site functionality.

See Appendix –K. For BIA Planning Factors

Key Organizational Communication Functions

Crisis Communication addresses how to communicate with employees and their families, the public, key customers, suppliers, stakeholders, management, and the media during a crisis.

The usual items addressed are:
- identification of official spokesperson(s)
- Press releases
- Press conferences and interviews

- Status reports
- Communication with local, state, and national emergency services, civil authorities, weather bureaus, etc.
- Communication with provincial / state and local governmental agencies (fire department, law enforcement, EMS and transportation)

Effective communication is indispensable before, during and after an emergency. An organization must skillfully communicate internally and externally to maintain maximum operational performance. Due to recent national and international disasters, organizations are becoming increasingly aware of the risk of serious business interruptions and the necessity of maintaining better communication with its personnel.

Business Continuity is not simply a matter of protecting data and information. It's also about contacting key business continuity team members, notifying on-call emergency responders, mobilizing and informing employees, keeping management abreast of developments and notifying customers and vendors of important information. It's the ability to communicate in a quick, efficient, and reliable manner in order to protect lives, prevent or limit economic loss and avoid miscommunication.

Internal Communication consists of internal networks positioned within the organization's operational perimeter, i.e. CPU, terminals, printers, connected by a range of methods to exchange data over short distances.

External Communication is a company's communication plan and notification system. It's a critical ingredient of the Business Continuity Planning. External communication companies such as MCI, Verizon and other external organizations must be an active part of your business continuity plan. Establish pre-disaster communication contracts and involve your communication service in your emergency exercises.

What are the internal emergency notification elements in your Business Continuity Planning?

* Building Evacuation – Notify specific floors, buildings or entire campuses when major disasters hit.

- Emergency Responders – Contact emergency response teams and public safety organizations, e.g. police, fire, EMT, Hazmat, etc.

- Weather Alerts – Relay vital communications related to hurricanes, tornadoes, floods, lightning, wildfires, etc.

- Blackouts – Send critical notifications even when electricity, landlines and computers are down.

- Terrorism – Alert employees to changing terrorist threat warning levels from Homeland Security.

- Personnel Scheduling – Mobilize additional workers to meet demand: shut down facilities; reschedule or cancel shifts.

- Remote Roll-Calling – Verify safety & location of employees with active feedback feature (Press 1 if at home; Press 2 if off-site)

- Routine Notification – Deliver routine office messages, contact business continuity teams, and security personnel.

- Phone Tree Rosters – Physically call each person in your organization.

Key employees during a disaster must be identified and clearly know their responsibilities. Employees must, without a doubt, understand their positions and roles during an emergency. Employees who occupy key positions must be aware of the possible working conditions that may result from a disaster to include extended hours.

Some of these positions are as follows:

Emergency Essential

They are vital in the event of an emergency; the survivability of the business depends on their skills and experiences. (Senior, facility engineer)

Key Personnel

Management is the key factor in every organization. Whether it is for decision making, either delegated or appointed managers are required during an emergency. (Vital data administrators and system operators)

Mission Essential

They are required to produce or to service as the core element to develop or maintain business continuity. (Machine operators)

Not Critical

Home and Tele-commuters

The "Call Tree" is the traditional alert notification process used by the military and corporate America. This is a process where all of your employee's have provide contact phone numbers.

The most effective process is for each person to call the next chronological person on the telephone roster until he/she physically speaks to someone. Every person in your organization will "call-in", acknowledging that they received the alert notification. Furthermore, an assigned person will continue to call personnel that have not responded. The "Call Tree" will list key personnel and all possible contact phone numbers:

NAME	WORK	HOME	CELL
Kenneth Lionel	703-XXX-XXXX	301-XXX-XXXX	202-XXX-XXXX

Chapter 3

The Federal Government is Mandated by Regulations and the (FPC 65) to provide sound principles for COOP activities

Continuity of Operations (COOP) Lessons Learned From Hurricanes Katrina, Rita, and Wilma.

Office of National Security Coordination (FEMA)
February 2006

Hurricanes Katrina, Rita, and Wilma represented the largest Federal Continuity of Operations (COOP) event in the history of the nation:

Greater than September 11, 2001
Greater than the 2003 Northeast blackout
Greater than the 1989 Loma Prieta earthquake

The purpose was to capture the issues, best practices and lessons learned by Federal Departments and Agencies relating to their COOP activities.

The goals of this report were:

To identify systemic operational issues that impacted the ability to execute their essential functions
To identify root causes of the issues
To build consensus on issues that need to be raised
To identify lessons learned – positive and negative – that should be shared

This "Lessons Learned" document was distributed to other Departments and Agencies across the nation. The document focused on:

Actions to take before an event
Actions to take during an event
Actions to take after an event
Coordination with other government entities

The Discussion Focused on: FPC 65 or "11 Elements of a Viable COOP Plan"

Plans & Procedures
Essential Functions
Vital Records and Databases
Devolution of Control & Direction
Reconstitution
Tests, Training & Exercises
Orders of Succession
Delegations of Authority
Alternate Operating Facilities
Human Capital
Interoperable Communications

Lessons Learned – Before an Event

The Department and Agency COOP Plan must be updated at least annually
Plan COOP for up to 30 days when activated.
Identify and coordinate for alternate facilities and reconstitution spaces before an event occurs

Activate early and anticipate evacuation delays
Identify key personnel and practice cascade call-downs

Develop and implement a good Vital Records program

Best Practices – Before an Event

Pre-identify where to send Emergency Relocation Group (ERG) members and their families before an evacuation

Provide all employees with agency ID cards with emergency contact information on reverse

Establish a central call-in telephone number, toll-free if possible

Store vital records on electronic media devices and on a network drive for off-site accessibility

Create a <u>Devolution of Operations Plan</u> and practice it

<u>The Author's Comments are underlined; furthermore, the Author's comments are provided for additional support for plan development:</u>

Before the Disaster
<u>Warnings: Raised Alert Levels & Situation Monitoring</u>

<u>Planning:</u>

- *<u>Management has copies of your organization's BIA/MEF available.</u>*
- *<u>Have a critical decision maker (Chain of Command) in place</u>*
  <u>Warning/No Warning – Duty/Non-Duty</u>
  <u>Stand-down – Transition Back to Normal</u>
- *<u>Delegations of Authority & Orders of Succession</u>*
- *<u>Continuity Staff Alert/Notification/Movement/Training</u>*
- *<u>Logistics Support, have a site guide for all alternate locations</u>*
- *<u>Alternate Sites, be prepared in activate, receive personnel and operate</u>*
- *<u>Have guidance on Storage/Protection/Availability of Vital Record/Material/Databases</u>*

It is critical to have a plan; your plan must be a "working document." All essential activities related to your organization must be addressed in order to keep your plans current. Pre-identify your organization's Emergency Coordinator. Emergency coordinators must stay abreast on current global and national emergency situations. The COOP Plans/BCP must be reviewed and updated monthly as systems and personnel changes. The plan will also require modifications. (This is a full-time job)

An unsighted factor is employees' family members. This issue is critical. It's human nature, family comes first. Employees must address family life support issues and support during planning. Activating and relocating depends on the magnitude of the disaster; your plan should address different options. Identify and coordinate for alternate/reconstitution facilities before an event occurs. (See Alternate Site and Hot, Warm, and Cold Site Concepts)

Identify key personnel and practice cascade call-downs. Have a communication plan/Call Tree and assign Key personnel. Develop and implement a good Vital Records program, replicating and/or backup vital data at all alternate locations. Pre-position supplies and logistical support.

Lessons Learned – During an Event

It is too late to practice activation and evacuation planning during an event
Keep supplies on-hand in the building in case of Shelter- in-Place situations
Keep in close contact with immediate supervisors.

Best Practices – During an Event

Set up lodging arrangements on-site for employees who are required to stay in the area for continuity
Alert devolution of operations contacts about possible transition
Set up "People Cells" that have full agency rosters to contact all employees to determine their status

The Author's Comments:

During the Disaster
Incident Priorities; Safety/Life Saving & Evacuation

- *100% Accountability of personnel and sensitive equipment*

- *Use Emergency Relocation Sites*

- *Situation Tracking & Reporting Immediate Response*

- *Activate Incident Priorities; MEF/BIA*

During and After the Disaster
Maintain Command and Control (C2)

- *4 Emergency Activities/ Centers*
    - *Command Center-Senior Management*
    - *Emergency Operations Center-Recovery & MEF/BIA*
    - *Incident Command Post-The Incident Site*
    - *Devolution Site-Last Hope for Command & Control*

Lessons Learned – After an Event

Prepare for logistical needs/shortages during the Response Level of an event.

Plan for difficulties in getting employees and vehicles back into the disaster area.

It is difficult coordinating the needs of Federal Departments and Agencies with the State and local governments after the event. Coordinate and plan with them ahead of time.

Use the plan! Focus on saving lives, safety, and security. Have alternate means of communicating and/or designated meeting locations.

Best Practices – After an Event

Once computers were shutdown at the primary site, many D/As could not link to them from their alternate facilities. Need to check ahead of time the organization's interoperable communication systems
Consider removing employees who live in affected areas from response teams
Identify supply sites (fuel, food, and water) in affected areas for Federal employees and civilian populations

The Author's Comments:

After the Disaster
Reconstitution, Relocate, and/or Rebuild

- *Employee's Responsibilities*
    - *Accountable*
    - *Contactable*
    - *Available*
- *Supervisor Responsibilities*
    - *Accountability*
    - *Guidance*
    - *Reporting*
- *Agency Responsibilities, Damage Assessment "Return to Normal" Response and Recovery*

Use the plan! Focus on saving lives, safety and security. Activate your organization's Alternate Sites and deploy Damage Assessment Team.

***The Devolution of Control & Direction site** is your last hope for operational and political control. During Hurricane Katrina, for approximately three days, there was no clear control and/or directions.*

The Federal Preparedness Circular (65) provides guidance to Federal departments and agencies for use in developing practical and executable contingency plans for continuity of operations (COOP). COOP planning facilitates the performance of the agency's essential functions during any emergency or situation that may disrupt normal operations. The United States Government policy is to have in place a comprehensive and effective program to ensure continuity of essential Federal functions continued under all circumstances. All Federal agencies shall have in place a viable COOP capability that ensures the performance of their essential functions during any emergency or situation that may disrupt normal operations.

COOP planning is merely a *"good business practice"* – part of the fundamental mission of agencies as responsible and reliable public institutions. For years, COOP planning had been an individual agency responsibility primarily in response to emergencies within the confines of the organization. The content and structure of COOP plans, operational standards, and interagency coordination, if any, were left to the discretion of the agency.

The changing threat environment and recent emergencies, including localized acts of nature, accidents, technological emergencies, and military or terrorist attack-related incidents, has shifted awareness to the need for COOP capabilities that enable agencies to continue their essential functions across a broad spectrum of emergencies.

Also, the potential for terrorist use of weapons of mass destruction has emphasized the need to provide the President a capability that ensures continuity of essential government functions across the Federal Executive Branch. COOP planning is an effort to assure that the capability exists to continue essential agency functions across a wide range of potential emergencies.

The objectives of a COOP plan include:

- Ensuring the performance of an agency's essential functions/operations during an emergency
- Reducing loss of life, minimizing damage and losses
- Realizing, as required, successful succession to office with accompanying authorities that are required in the event a disruption renders agency leadership
- Realizing, as required, successful succession to office with accompanying authorities that are required in the event a disruption renders agency leadership unable, unavailable, or incapable of assuming and performing their authorities and responsibilities of office
- Reducing or mitigating disruptions to operations
- Ensuring that agencies have facilities from which to continue to perform their essential functions/operations during an emergency
- Protecting essential facilities, equipment, records, and other assets
- Achieving a timely and orderly recovery from an emergency and resumption of critical functions to both internal and external clients during COOP
- Achieving a timely and orderly reconstitution from an emergency and resumption of full service to both internal and external clients
- Maintaining a test, training, and exercise program to support the implementation and validation of COOP plans

PLANNING CONSIDERATIONS:

In accordance with current guidance, a viable COOP capability:
- Must be capable of implementation both with and without warning
- Must be operational no later than 12 hours after activation
- Must be capable of maintaining sustained operations for up to 30 days
- Should include regularly scheduled testing, training, and exercising of agency personnel, equipment, systems, processes, and procedures used to support the agency during a COOP event
- Should provide for a regular risk analysis of current alternate operating facility

- Should locate alternate facilities in areas where the ability to initiate, maintain, and terminate operations is maximized
- Should take maximum advantage of existing agency field infrastructures and give consideration to other options, such as telecommuting locations, work-at-home, virtual offices, and joint or shared facilities
- Should consider the distance of alternate facilities from the primary facility and from the threat of any other facilities/locations (e.g. nuclear power plants or areas subject to frequent natural disasters)
- Agencies should develop and maintain their COOP capabilities using a multi-year strategy and program management plan. The plan should outline the process the agency will follow to designate essential functions and resources, define short and long-term COOP goals and objectives, forecast budgetary requirements, anticipate and address issues and potential obstacles, and establish planning milestones

ELEMENTS OF A VIABLE COOP CAPABILITY

At a minimum, all agency COOP capabilities shall encompass the following elements:

a. PLANS AND PROCEDURES: A COOP plan shall be developed and documented that when implemented will provide for continued performance of essential Federal functions under all circumstances. At a minimum, the plan should:
- Delineate essential functions and activities
- Establish orders of succession to key agency positions and a roster(s) of fully equipped and trained COOP essential personnel with the authority to perform essential functions and activities
- Provide for the identification and preparation of alternate facilities for relocated operations
- Outline a decision process for determining appropriate actions in implementing COOP plans and procedures
- Provide for attaining operational capability within 12 hours of activation

- Establish reliable processes and procedures to acquire resources necessary to continue essential functions and sustain operations for up to 30 days
- Provide for the ability to coordinate activities with personnel not deployed
- Provide for reconstitution of agency capabilities and transition from continuity operations to normal operations

b. **ESSENTIAL FUNCTIONS** All agencies should identify their essential functions as the basis for COOP planning. Essential functions are those functions that enable Federal Executive Branch agencies to provide vital services, exercise civil authority, maintain the safety and well being of the general populace, and sustain the industrial/economic base in an emergency. In identifying essential functions, agencies should:

(1) Identify functions performed by the agency, then determine which must be continued under all circumstances

(2) Prioritize these essential functions

(3) Establish staffing and resources requirements needed to perform these functions: information technology and telecommunications (ITT), identify hardware, identify consumable office supplies, and identify mission critical data and systems

(4) Integrate supporting activities to ensure that essential functions can be performed as efficiently as possible during emergency relocation

(5) Recognize that prioritization of functions can be situated on the dependent and agencies should prioritize functions against the most probable scenarios

(6) Defer functions not deemed immediately essential to agency needs until additional personnel and resources become available.

c. **DELEGATIONS OF AUTHORITY** To ensure rapid response to any emergency situation requiring COOP plan implementation, agencies should pre-delegate authorities for making policy determinations and decisions at headquarters, field levels, and other organizational locations, as appropriate? Generally, pre-determined delegations of authority will take effect when normal channels of direction are disrupted and terminate when these channels have resumed.

d. **ORDERS OF SUCCESSION** Agencies are responsible for establishing, promulgating and maintaining orders of succession to key positions. Such orders of succession are an essential part of an agency's COOP plan. Orders should be of sufficient depth to ensure the agency's ability to perform essential functions while remaining a viable part of the Federal government through any emergency.

(1) Include the orders of succession in the vital records of the agency

(2) Revise orders of succession as necessary and distribute revised versions promptly as changes occur

(3) Establish the rules and procedures designated to officials that they are to follow when facing the issues of succession to office in relation to any emergency situation

(4) Include in succession procedures the conditions under which succession will take place, method of notification, and any temporal, geographical, or organizational limitations of authority

e. **ALTERNATE FACILITIES** All agencies shall designate alternate operating facilities as part of their COOP plans, and prepare their personnel for the possibility of unannounced relocation of essential functions and/or COOP essential personnel to these facilities. Facilities may be identified from existing agency local/field infrastructures or external sources. Facilities should be capable of supporting operations in a low threat environment, as determined by the geographical location of the facility, a favorable assessment of the local threat, and/or the collective protective characteristics of the facility. In acquiring and equipping such facilities, agencies are encouraged to consider cooperative interagency agreements and to promote sharing of identified alternate facilities. However, care should be taken to ensure that shared facilities are not overcommitted during a COOP situation. Alternate facilities should provide:

(1) Sufficient space and equipment to sustain the relocating agency

(2) Capability to perform essential functions within 12 hours of activation and for up to 30 days under various threat conditions, including threats involving weapons of mass destruction

(3) Reliable logistical support, services, and infrastructure systems

(4) Consideration for the health, safety, security, and emotional well being of relocated employees

(5) Interoperable communications with all identified essential internal and external organizations, customers, and the public

(6) Computer equipment, software, and other automated data processing equipment necessary to carry out essential functions. (Annex E Alternate Facilities)

f. **INTEROPERABLE COMMUNICATIONS** Execution of essential government functions requires a high assurance of connectivity between key agency personnel during crisis, disasters, or wartime conditions. Elements of a viable interoperable communications program include:

(1) Capability commensurate with an agency's essential functions and activities

(2) Ability to communicate with contingency staffs, COOP essential personnel, leadership, and other agency components as applicable

(3) Ability to communicate with other agencies, organizations, customers, and COOP essential personnel, as applicable

(4) Access to data and systems necessary to conduct essential activities and functions

(5) Communications systems for use in COOP implementation both with and without warning

(6) Must be available to support COOP operational requirements

(7) Must provide sustained operations for up to 30 days

(8) Should provide interoperability with existing field infrastructures

g. **VITAL RECORDS AND DATABASES** The identification, protection, and ready availability of electronic and hardcopy documents, references, records, and information systems needed to support essential functions under the full spectrum of emergencies is another critical element of a successful COOP plan. Agency personnel must have access to and be able to use these records and systems in conducting their essential functions. Categories of these types of records may include:

Emergency Operating Records Vital records, regardless of media, essential to the continued functioning or reconstitution of an agency during and after an emergency. Included are emergency plans and directives, orders of succession, delegations of authority, staffing assignments, and related records of a policy or procedural nature that provide agency essential personnel with guidance and information resources necessary for conducting operations during a COOP situation, and for resuming normal operations at its conclusion.

Legal and Financial Records These include vital records, regardless of media, critical to carrying out an agency's essential legal and financial functions, and protecting the legal and financial rights of individuals directly affected by its activities. Included are records having such value that their loss would significantly impair the execution of essential agency functions, to the detriment of the legal or financial rights, and/or entitlements of the agency of the effected individual(s)? Examples of this category of vital records are accounts receivable; contracting and acquisition files; official personnel records; Social Security, payroll, retirement, and insurance records; and property management and inventory records.

The COOP Plan should account for the identification and protection of the vital records, systems, and data management software and equipment, to include classified or other sensitive data, as applicable, necessary to perform essential functions, and to reconstitute normal agency operations after the emergency. To the extent possible, agencies should preposition and update on a regular basis, duplicate records or back-up electronic files.)

h. **TEST, TRAINING, AND EXERCISES** Testing, training, and exercising of COOP capabilities are essential to demonstrating and improving the ability of agencies to execute their COOP plans. Tests and exercises serve to validate, or identify for a subsequent corrective action program, specific aspects of COOP plans, policies, procedures, systems, and facilities used in response to an emergency situation. Training familiarizes COOP essential personnel with the functions they may have to perform in a COOP situation. All agencies shall plan and conduct tests, training, and exercises to demonstrate viability and interoperability of COOP plans. COOP test, training, and exercise plans should provide:

Individual and team training of agency COOP essential and emergency personnel to ensure currency of knowledge and integration of skills necessary to implement COOP plans and carry out essential functions. Team training should be conducted at least annually for COOP essential personnel on their respective COOP responsibilities.

Internal agency testing and exercising of agency COOP plans and procedures to ensure the ability to perform essential functions and operate from designated alternate facilities and other COOP environments. This testing and exercising shall occur at least annually.

Testing of alert and notifications procedures and systems for any type of emergency at least quarterly.

Refresher orientation for COOP essential personnel arriving at alternate operating facilities. The orientation should cover the support services available at the facility, including communications and information systems for communicating if the normal operating facility is still functional; and administrative matters, including supervision, security, and personnel policies.
Joint agency exercising of COOP plans, where applicable and feasible.

Periodic testing to ensure that equipment and procedures are maintained in a constant state of readiness.

i. **DEVOLUTION OF CONTROL AND DIRECTION** Devolution planning supports overall COOP planning and addresses catastrophic disasters rendering an agency's headquarters incapable of supporting the execution of its essential functions. The devolution option of COOP shall be developed to address how an agency will identify and conduct its essential functions in the aftermath of a worst-case emergency. At a minimum, the plan should:

(1) Identify prioritized essential functions and determine necessary resources to facilitate their immediate and seamless transfer to a devolution site

(2) Roster fully equipped and trained personnel at the designated devolution site with the authority to perform essential functions and activities when the devolution option of COOP is activated

(3) Identify the likely triggers that would initiate or activate the devolution option

(4) Specify how and when direction and control of agency operations will be transferred to the alternate operating facility (ies)

(5) List necessary resources (people, equipment, and materials) to facilitate the ability to perform essential functions at the devolution site

(6) Establish reliable processes and procedures to acquire resources necessary to continue essential functions and sustain operations for extended periods

(7) Establish capabilities to restore or reconstitute agency authorities to their pre-event status upon termination of devolution

j. **RECONSTITUTION** Extensive coordination is necessary to procure a new facility and staffing once an agency suffers the operational loss of its originating facility due to an event directly affecting the facility or collateral damage from a disaster rendering the structure unavailable for reoccupation. Agencies shall identify and outline a plan to return to normal operations once an Agency Head or their successor determines reconstitution operations can begin.

k. **HUMAN CAPITAL.** COOP human capital planning and preparedness encompasses the following areas:

(1) Agency planning and readiness;

(2) Designation of emergency employees and other special categories of employees;

(3) Dismissal or closure procedures;

(4) OPM and media announcements on government operating status;

(5) Status of non-emergency employees and non-special categories of employees;

(6) Sample agency guidelines for communicating to employees;

(7) Methods of employee communications,

(8) Employee awareness of changes in building operations,

(9) Pay flexibilities,

(10) Staffing flexibilities,

(11) Benefit issues,

(12) Employee roles and responsibilities

l. COOP IMPLEMENTATION Relocation may be required to accommodate a variety of emergency scenarios. Examples include scenarios in which:

An agency headquarters is unavailable and operations can shift to a regional or field location. A single agency facility is temporarily unavailable and the agency can share one of its own facilities or that of another agency.

Many, if not all, agencies must evacuate the immediate area.

While any of these scenarios involves unavailability of a facility, the distinction must be made between a situation requiring evacuation only and one dictating the need to implement COOP plans. A COOP plan includes the deliberate and pre-planned movement of selected key principles and supporting staff to a relocation facility. As an example, a sudden emergency, such as a fire or hazardous materials incident, may require the evacuation of an agency building with little or no advance notice, but for only a short duration. Alternatively, an emergency so severe that an agency facility is rendered unusable and likely will be for a period long enough to significantly impact normal operations, may require COOP plan implementation. Agencies should develop an executive decision process that would allow for a review of the emergency and determination the best course of action for response and recovery. This will preclude premature or inappropriate activation of an agency COOP plan. One approach to ensuring a logical sequence of events in implementing a COOP plan is time phasing. A suggested time-leveled approach for COOP activation and relocation, alternate facility operations, and reconstitution follows:

LEVEL 1 – ACTIVATION AND RELOCATION (0-12 HOURS)

Notify alternate facilities managers of impending activation and actual relocation requirements.

Notify the FEMA Operations Center and other appropriate agencies of the decision to relocate and the time of execution or activation of call-down procedures.

Activate plans, procedures, and schedules to transfer activities, personnel, records, and equipment to alternate operating facility (ies).

Notify initial COOP contingency staff to relocate.

Instruct all other personnel on what they are to do.

Assemble necessary documents and equipment required to continue performance of essential operations at alternate operating facility (ies).

Order necessary equipment/supplies, if not already in place.

Transport documents and designated communications, automated data processing, and other equipment to the alternate operating facility (ies), if applicable.

Secure the normal operating facility physical plant and non-movable equipment and records, to the extent possible.

Continue essential operations at the normal operating facility, if available, until alternate facility (ies) is operational.

Advise alternate operating facility manager(s) on the status of follow-on personnel.

LEVEL II – ALTERNATE FACILITY OPERATIONS (12 HOURS)

Provide amplified guidance to COOP essential personnel and all other employees.

Identify replacements for missing personnel and request augmentation as necessary.

Commence full execution of essential operations at alternate operating facility (ies).

Notify the FOC and all other appropriate agencies immediately of the agency's alternate location, operational, and communications status, and anticipated duration of relocation, if known.

Develop plans and schedules to level down alternate facility (ies) operations and return activities, personnel, records, and equipment to the primary facility when appropriate.

LEVEL III RECONSTITUTION (TERMINATION AND RETURN TO NORMAL OPERATIONS)

Inform all personnel that the threat of an actual emergency no longer exists, and provide instructions for resumption of normal operations.

Supervise an orderly return to the normal operating facility, or movement to other temporary or permanent facilities using a leveled approach if conditions necessitate.

Report status of relocation to the FOC and other agency points of contact (POC), if applicable.

Conduct an after-action review of COOP operations and effectiveness of plans and procedures as soon as possible, identify areas for correction, and develop a remedial action plan.

I. *RESPONSIBILITIES* The following responsibilities should be clearly outlined in agency COOP planning guidance and internal documents:

Each Agency Head is responsible for:

(1) Appointing an agency COOP program POC

(2) Developing a COOP Multi-Year Strategy and Program Management Plan

(3) Developing, approving, and maintaining agency COOP plans and procedures for headquarters and all subordinate elements, which provide for: Identification of agency essential functions; pre-determined delegations of authority and orders of succession; contingency staffing to perform essential functions; alternate

operating facilities; interoperable communications, information processing systems, and equipment; and protection of vital records and systems

(4) Conducting tests and training of agency COOP plans, to include COOP contingency staffs, and essential systems and equipment, to ensure timely and reliable implementation of COOP plans and procedures

(5) Participating in periodic interagency COOP exercises to ensure effective interagency coordination and mutual support

(6) Notifying the FOC and other appropriate agencies upon implementation of COOP plans

(7) Coordinating intra-agency COOP efforts and initiatives with policies, plans, and activities related to terrorism under PDD-62 and Critical Infrastructure Protection under PDD-63

In addition, FEMA, as the Department of Homeland Security (DHS) Executive Agent is responsible for:

(1) Serving as the Executive Agent for Federal Executive Branch COOP

(2) Coordinating COOP activities of Federal Executive Branch agencies

(3) Issuing COOP guidance, in cooperation with the General Services Administration, to promote understanding of, and compliance with, the requirements and objectives of governing directives

(4) Chairing the COOP Working Group (CWG), which serves as the principal interagency forum for discussion of COOP matters such as policy guidance, plans, and procedures, and for dissemination of information to agencies for developing and improving their individual COOP plans

(5) Conducting periodic assessments of Executive Branch COOP capabilities and reporting the results to the National Security Council

FPC 65 Process Flow

Plans & Procedures
Essential Functions
───────────────────

Vital Records and Databases
───────────────────

Interoperable Communications
───────────────────

Orders of Succession
Delegations of Authority
───────────────────

Alternate Operating Facilities
───────────────────
Human Capital
───────────────────

Devolution of Control & Direction
Reconstitution
───────────────────

Tests, Training & Exercises
───────────────────

Chapter 4

Business Continuity Management

Business continuity is the process of developing advance arrangements and procedures that enable an organization to respond to an event in such a manner that Critical Business Functions continue without interruption or essential change.

Figure 7

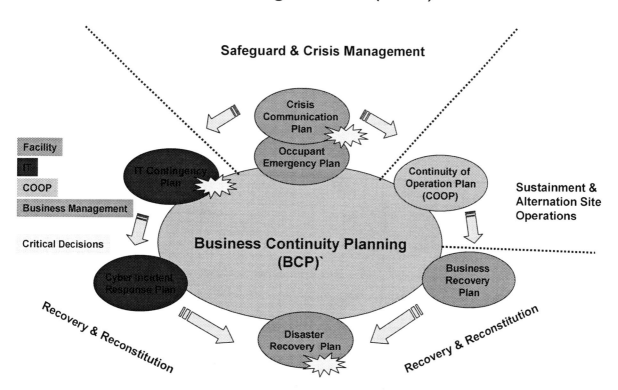

(Referenced from NIST Standard 800-34 and FPC-65)

Understanding Business Continuity Management "the Big Picture" (Planning)! What are the required plans?

During a crisis at any level, an organization must establish a Crisis Action Team (CAT) to address critical management and operational tasks. The Team is pre-established and members from key divisions in your organization must be assigned. The CAT Team provides an immediate means to tackle quick decisions that may affect the organization mission. Senior Management must delegate the proper authority and provide the CAT the means to take decisive action to resolve important problems. The CAT will initially alert, notify, and activate COOP procedures, if required.

Occupant Emergency Plan (OEP): The OEP provides the response procedures for occupants of a facility in the event of a situation posing a potential threat to the health and safety of personnel, the environment, or property. Such events would include a fire, hurricane, criminal attack, or a medical emergency. OEPs are developed at the facility Level, specific to the geographic location and structural design of the building.

See Annex?

Crisis Communications Plan (CCP): The Crisis Communications Plan addresses internal and external dissemination of information prior to and/or during a disaster/incident. The Crisis Communications Plan procedures should be coordinated with all other plans to ensure that only approved statements are released to the public.

The communications plan typically designates specific individuals as the only authority for answering questions from the public regarding disaster response. It may also include procedures for disseminating status reports to personnel and to the public. Templates for press releases are included in the plan.

Continuity of Operations Plan (COOP): The COOP focuses on restoring an organization's essential functions at an alternate site and performing those functions until returning to normal operations.

Business Recovery Plan (BRP): The BRP addresses the restoration of business functions after an emergency, but unlike the BCP, lacks procedures to ensure continuity of critical functions throughout an emergency or disruption. Development of the BRP should be coordinated with the disaster recovery plan and the BCP. The BRP may be appended to the BCP.

IT Contingency Plan: A process that focuses on data/computing center and/or local/wide area network recovery following a disruption including specific actions for restoring or recovering IT and other systems after they fail. These plans usually include the procedures to safeguard information by conducting backups or similar procedures to permit the restoration of information. These plans are prepared by system administrators, but include appropriate links to the business continuity plans of all functions that rely on that system or IT component. *Information Assurance (IA)* is Information operations that protect and defend information and information systems by ensuring their availability, integrity, authentication, confidentiality, and non-repudiation. This includes providing for restoration of information systems by incorporating protection, detection, and reaction capabilities.

Disaster Response Plans: Provide for immediate reaction after a disaster occurs. Response plans are directed toward the safety of people and company assets. Plans are required for evacuating the building and coping with other life/safety emergencies as required. Whether it is a fire, an explosion, a chemical spill, a tornado, flooding, or some other disasters, response plans are imperative for management, employees, and customers to address sound and life-threatening decisions during dangerous times. The safety of people has to be the top priority.

Disaster response plans, whether they focus on life-threatening issues or company assets, must be written as clear, "easy to use" instructions for the first response team to use as they begin to cope with an emergency.

Cyber Incident Response Plan: The Incident Response Plan establishes procedures to address cyber attacks against an organization's IT system(s). These procedures are designed to enable security personnel to identify, mitigate, and recover from malicious computer incidents, such as unauthorized access to a system or data, denial of service, or unauthorized changes to system hardware or software (e.g. malicious logic such as a virus, worm, or Trojan horse).

Continuity planners must understand the Risk and Control Mechanic of their environment

They must determine the events, environment surroundings, and possible effects on their organization and/or facilities. Their primary function is to prevent or minimize the effects of potential loss; furthermore, providing cost-benefit analysis to justify investments in controls mitigates risk.

- There are eight important factors to consider:
- Understanding the possible problems, probabilities and risk reduction/mitigation
- Identify Potential Risk to the Organization
- Identify Outside Expertise Required
- Identify Vulnerabilities/Threat/Exposure
- Identify Risk Reduction/Mitigation Alternative
- Identify Credible Information Sources
- Interface with Management to Establish Acceptable Risk Levels
- Document and Present Result

Continuity planning is planning done before a disaster to ensure that the business can resume functioning as soon as possible after the disaster. The success or failure of continuity planning has a direct effect on how quickly your business can begin to function after a disaster.

Continuity requires that key documents be available and current. Here are a few examples of key documentation that should be stored at an alternate site or outside the building and updated on a regular basis:

Business Continuity Plan

Lists of employees and contact information

Insurers and emergency agencies information

Back-up copies of vital data and documentation, if possible, on a back-up server to include the necessary software to run the files

Alert roster, account numbers, passwords, and telephone number of Key Personnel

Continuity plans need to be reviewed regularly as business procedures and processes change. The relationship between emergency response plans, continuity planning, and actual employee behavior must also be examined. Simply put, if the disaster response plan says that a recent copy of the server's files is available in a specific offsite location, it still must be assured someone is actually making the copies and sending them to offsite storage on schedule or that simultaneous back-up systems are working. It isn't enough to simply write it in the plan; it has to actually be happening.

Identify and describe Mission Essential Functions of the business.

Outline a decision process for determining appropriate actions in implementing continuity plans and procedures.

Establish delegations of authority for making policy determinations, decisions, and orders of succession for key positions to ensure the organization's ability to perform MEF during all contingencies, threats, and events. Such delegations shall be used for reconstituting the organization of leadership and control.

Include procedures for advisories, alerts/notification, and communication systems with instructions for relocation to pre-designated facilities, with or without warning, during duty and non-duty hours. Provide for attaining operational capability within established time periods, according to the Business Mission requirements.

Identify staff prepared to relocate to secure facilities, maintain command and control, execute MEF, reorganize and redirect resources, conduct required interagency coordination, and implement decisions.

Establish reliable processes and procedures to acquire resources necessary to continue MEF and sustain operations according to the Company mission requirements.

Before a disaster, know what types of insurance forms would need to be filled out to substantiate a claim. Learn about State and Federal disaster relief programs and the documentation required.

Continuity Planning includes:

Business Continuity Planning (BCP)
IT Disaster Recovery Planning
Emergency Planning
Disaster Planning

Other types of planning activities and programs that shall be integrated into Continuity Planning include:

Critical Infrastructure Protection
Information Assurance
Mitigation Plan ***

Mitigation is action taken before a disaster to reduce the risk of disaster or reduce the impact of the event. A mitigation plan is a plan to minimize damage and/or prevent an emergency from becoming a disaster by taking specific actions before an emergency occurs. Requiring more time and money does not make an action impossible. It simply means that you also have to plan how to get the money. Increasingly, agencies such as the Federal Emergency Management Agency (FEMA) are promoting mitigation because they have done the math. They know that mitigation saves money in the long run. They also know that mitigation can save lives.

Mitigation activities cannot be done to help your business until there is a plan that details what should be done as mitigation. A mitigation plan is a list of mitigation activities with priorities, commitments, and a timeline. Once a mitigation plan is written, it should be reviewed regularly and evaluate how well mitigation goals are being met. Mitigation Planning is part of the BCP.

Responsibilities of the Continuity Manager (CM):

The CM should provide guidance to and oversee the company's continuity planning program. The Continuity Manager should:

Prepare the Continuity Plan and ensure supporting contingency plans for the company are prepared, coordinated, tested, and updated.

Assess the overall status of Company's continuity planning.

Recommends courses of action to correct deficiencies.

Receive, review, and maintain current editions of all Company Continuity Plans and notify Senior Management of plans that are not in compliance or are in potential conflict with other plans.

Identify means that may enable the Company to take advantage of parallel processes in organizations and disseminate that information as needed.
Coordinate other company sites to discuss issues of interest that may support the program.

Be the Company's coordinator and representative for external organization related to BCP and related policies and programs.

Direct and conduct periodical tests and assessment for the BCP, no less than annually.

Verify that the BCP complies with the latest company policies and procedures.

Advocate resource and funding alternatives.

Understand Senior Management's point of view during Continuity Program meeting.

Your Organization's Business Continuity Plan should include ...

Preventive Controls:

Identify controls that are in place in order to deter, detect, and reduce impacts to business operations and supporting IT. These controls may address facility access, personnel access and security, IT infrastructure, physical infrastructure, and data confidentiality, integrity, and availability.

Incident Response:

Incident Response addresses: *(for example)*
The process used to determine the nature of the incident
The process used to assess damage
The process used to differentiate between an interruption or a major disruption (escalation criteria) the process used to declare an event/incident. (Who? Delegation of Authority, Devolution of Control, and Order of Succession)

Business Operations:

Actions are identified that must be taken based upon the impact of the incident: *(for example)*
What are the essential and less time critical business functions that should be restored and in what order?
What is the point of the information recovery?
What are the responsibilities of your staff?
What are your vital records and equipment needs?
What activities should be performed for resumption and recovery?
Identify restoration activities at the original site or new location?

Information Technology Recovery:

Addresses the procedures and capabilities required in order to recover all supporting information technology such as: *(for example)*

Management of IT Operations
Administrative/Logistics
Application Support
Data Preparation
Database Systems
Technical Services
Network Engineering
Support Services

Addresses how to recover support functions such as transportation, equipment and supply purchasing, legal needs, security, and personnel/human resource services.

Related issues may include, but are not limited to: *(for example)*

Contracts
Facilities management
Pay and staffing
Procurement
Travel authorizations

Plan Testing and Implementation:

The following questions are answered:
How will the plan be tested?
How will the test results be recorded?
What is the schedule for testing?
What criteria will be used to evaluate the test results?

Crisis Communication Section:

Addresses how to communicate with employees and their families, the public, key customers, suppliers, stakeholders, management, and the media during a crisis. Issues addressed are:

Identification of official spokesperson(s)

Press releases

Press conferences and interviews

Status reports

Communication with local, state, and national emergency services, civil authorities, weather bureaus, etc.

Communication with governmental agencies including provincial/state, and local (fire department, law enforcement, EMS, and transportation)

Three Phase BCP:

1. Project Conception Period

Establish Control measures

Establish a Steering committee

A Trained Business Contingency Planner or the Site Emergency Manager should lead and advise on all matters concerning the plan

Middle Manager is key during this process; assign Manager's key roles

Team leaders are key players on the steering committee, i.e. assessment team, recovery team, etc.

The Steering Committee should draft a BCP Thought Process of agree on a concept

Business Continuity Scope

Objectives

Assumptions

Cost

2. Requirement Period

Risk Analysis

Fact Finding

Control Measures (Fail Mode Affect Analysis)

Alternate Strategies

Cost-Benefit Analysis

Budget

3. Developmental Approach Period

(1) SITUATION:

    a. Scope
    b. Objective
    c. External Organizations and Agencies: Impacts to the organization

(2) MISSION:

    a. Business Recovery Organization, responsibility

(3) EXECUTION:

    a. Concept of Operation: Scenario of Plan and Composition
    b. Duties and Responsibilities: Recovery Teams Concept

(4) SERVICE AND SUPPORT:

    a. Personnel Control Program
    b. Vital Records and Off-site Storage Program
    c. Operational Hot and Cold Site

(5) COMMAND AND SIGNAL:

    a. Escalation
    b. Notification
    c. Plan Activation

(4) Execution Period:

    a. Evacuation Procedure
    b. Command and Control of ICC, EOC
    c. Delegation of Authority
    d. Detailed Re-establishment Procedures

        e.   Contracts and Purchase of Recovery Resource

(5) **Analysis Processes and Updating Period:**

        a.   Actual Exercises
        b.   Plan Evaluation
        c.   Training and Awareness
        d.   Budget Review and Update
        e.   Reporting and Audits
        f.   Plan Distribution and Security

Thought Process   Figure 8

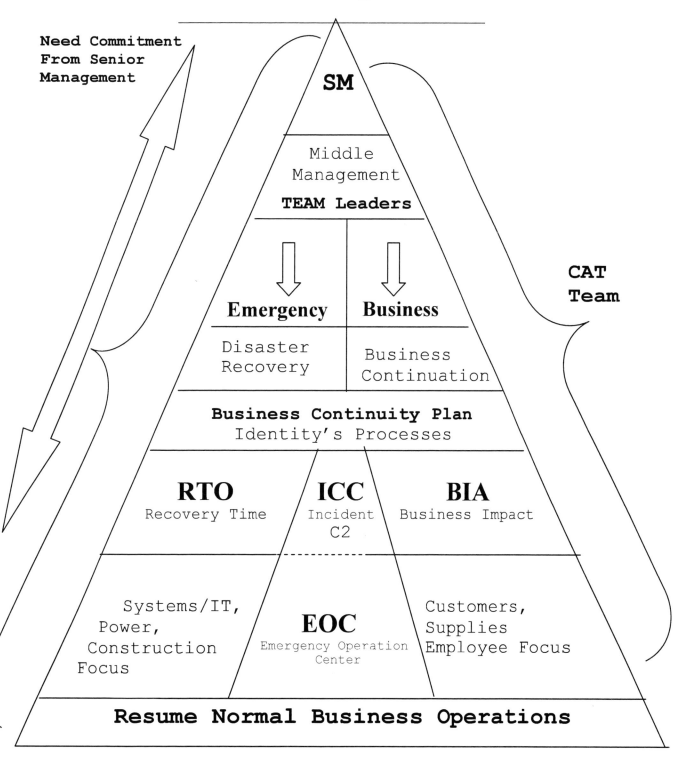

Senior Management Functions:

Senior management is responsible for all organization's functions and must clearly understand "Business Continuity Management" through a CAT Team or other means. Likewise, a middle management level or director shall be responsible for each department's segment of its BCP.

Middle Management's Functions: Support and direct the Business Impact Analysis (BIA) and the Mission Essential Functions (MEF) processes.

There are two primary focuses:

1. Disaster Recovery
2. Business Continuation

During a disaster, middle management must be directly involved in the Command and Control and keep senior management abreast of important issues and critical decisions.

The Incident Command Center (ICC) will forward information to the EOC. The EOC will track issues and inform senior management of critical decisions that are required. Management must understand the disaster impact on the business and aggressively mitigate possible problems. (BIA) Management must know "Recovery Times Points" in advance, the organization's Recovery Time Objective, are critical factor in providing continuous and uninterrupted business to your customers.

Recovery Focus –RTO and Business Focus BIA should be managed by different managers totally focusing on their area.

Chapter 5

Security & Business Continuity Management

Security is "TOP Driven." A superior security program is based on having skilled and capable "Security Managers." In the past, an ideal security professional was a retired police officer, prior military or just a young, wanna-be stub. This perspective must change; the quality of your security staff may be the difference in life or death. Disaster management is directly related to security.

Hurricane Katrina is a good example. Was there security or protection in the Dome? Security is the first line of defense for your organization's BCM program. Continuity Management includes securing your faculties and providing your personnel the best possible protection. Having sound security procedures, testing and exercising your security personnel, will make a difference during a disaster. In the event of a disaster, your security personnel are your organization's "first responders." Will they know how to perform critical functions prior to your emergency coordinator arrival?

Security is involved in the Continuity of Operations Plan (COOP): The COOP focuses on restoring an organization's essential functions and securing the Incident Site and performing those functions until returning to normal operations. Occupant Emergency Plan (OEP): The OEP provides the response procedures for occupants of a facility in the event of a situation posing a potential threat to the health and safety of personnel, the environment, or property. Such events would include a fire, hurricane, criminal attack, or a medical emergency. OEPs are developed at the facility level, specific to the geographic location and structural design of the building.

IT Contingency Plan: *Information Assurance (IA)* is information operations that protect and defend information and information systems by ensuring their availability, integrity, authentication, confidentiality, and non-repudiation. This includes providing for restoration of information systems by incorporating protection, detection, and reaction capabilities. Disaster response plans provide for immediate reaction after a disaster occurs. Response plans are directed toward the safety of people and company assets. Plans are required for evacuating the building and coping with other life/safety emergencies as required. Whether it is a fire, an explosion, a chemical spill, a tornado, flooding, or some other disaster, response plans are imperative for management, employees and customers to address sound and life-threatening decisions during dangerous times. The safety of people has to be the top priority. Disaster response plans, whether they focus on life-threatening issues or company assets, must be written as clear, "easy to use" instructions for the first response team to use as they begin to cope with an emergency.

Cyber Incident Response Plan: The Incident Response Plan establishes procedures to address cyber attacks against an organization's IT system(s). These procedures are designed to enable security personnel to identify, mitigate, and recover from malicious computer incidents, such as unauthorized access to a system or data, denial of service, or unauthorized changes to system hardware or software (e.g., malicious logic such as a virus, worm, or Trojan horse).

We must also consider the "Criminal Elements." Criminals are also evacuated. "Hurricane Katrina" is an example. Furthermore, people, to conclude, criminals will normally do what's necessary to survive. This element effects all disasters, not just Katrina! DETERRENCE (Is the Key)

1. Desire

2. Means (weapons and tools)

3. Opportunity

Criminals and belligerents usually have both the desire and the means to commit a crime and/or carry out an aggressive act of violence; they may or may not have the opportunity. Sometimes they only have the opportunity if we provide it for them. It is paramount, therefore, that site selection and the security measures employed, work to our advantage in reducing the number of opportunities these elements may find to violate our personnel and/or property.

Figure 9

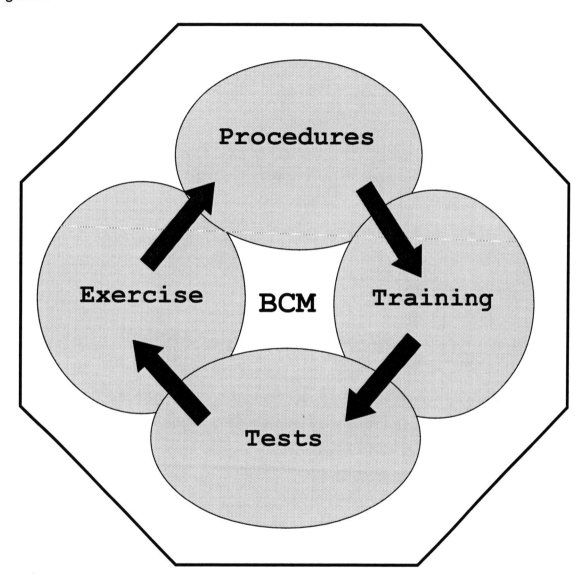

Security & Business Continuity Management

Your organization's BCM program is the Big Picture. You must have Sound and validated Procedures "in place." Your personnel must Train on a monthly basis, Test on the bi-monthly basis and Exercise on a Quarterly basis. Fig 9 displays an annual process considering personnel turnover and documentation up-keep.

PROCEDURAL BOUNDARIES

A. Physical boundaries are complimented by procedures, access to certain areas, who has access to what and under what circumstances, and when.

B. Controlled access to the building, accomplished by posting a guard or by simply having a receptionist in the front lobby to screen people entering the building.

C. Record keeping system of persons, who enter controlled or sensitive areas, is a low profile security measure. Record keeping systems should be utilized for Inventory control at warehouses, stores, and distribution sites. Visitor records. Deliveries of materials and supplies. Key control.

D. A procedure requiring "two" staff members to do nighttime security checks.

E. Procedures to check doors and windows prior to closing.

Good office security procedures are paramount to securing property and personnel. The security manager will manage the security program, in conjunction with overseeing the entire security overview of your organization/facilities to include inspection and maintenance of security systems and devices.

These activities are many to include areas listed:

A. Key control

All keys will be strictly controlled.
Keys must be kept on your person at all times, not left in drawers or on desktops.
Key control systems are imperative to determine:

Access to certain buildings or areas.

How and where the keys are kept.

How many keys are outstanding and who has them.

The necessity for re-keying locks.

Extra keys will be kept in a locked key control box and identified by code only. All lost or stolen keys will be reported immediately and all locks re-keyed. Staff who loses keys may be required to incur the cost of re-keying all locks. Keys must be collected immediately from any employee leaving the area or in the process of voluntary/involuntary termination.

B. Maintenance and Monitoring

Maintenance and monitoring of security devices on doors, windows, and within the building will be performed. The security and disaster management personnel will identify and inspect all areas and equipment that may cause or be subject to a disaster. These areas will include:

a. Building structure
b. Grounds
c. HVAC system
d. Electrical appliances and wiring
e. Plumbing and drainage
f. Housekeeping

C. Fire Safety

The security and disaster management personnel will manage the fire safety program.

This will include inspection and maintenance of fire protection systems and devices. To include:

a. Fire extinguishers

b. Fire alarm system

c. Smoke and heat detectors

d. Fire suppression system (sprinklers, etc.)

e. Liaison with the Fire Department

f. Staff training

D. Storage Areas

The security and disaster management personnel will ensure periodic inspection of collection storage areas with attention to:

a. Signs of leaks, water damage, etc.

b. Signs of mold, insect, or rodent infestation

c. Fire hazards

Inspections will include any offsite storage areas used for the collection.

E. Computer Backups

An important element of disaster mitigation is routine backup and offsite storage of computer records. To the extent that originals or duplicates are held elsewhere, the organization's vulnerability to disaster is reduced.

Information about computer backups and offsite storage of computer records will be provided into Business Continuity Plans.

F. Physical Boundaries:

Access to a site is controlled through a series of physical and procedural "boundaries," both designed as methods for blocking would-be intruders from gaining access to the site. These boundaries may be physical, such as walls or fences; physical and psychological, such as hedges or flowerbeds; procedural, such as security checkpoints; or any combination of these. Keep in mind these different boundaries when selecting a site. Clearly defined perimeters, even if just psychological, play a major role in discouraging intruders.

Check for natural boundaries, such as hedges or tree lines that separate the outer perimeter from the roadway or fields.

Check to see if inner perimeter boundaries exist. Inner perimeter boundaries may be hedgerows inside the line of trees that skirt the property.

a. PERIMETER SECURITY

1. Outer perimeter boundaries include:
Fences
Walls
Bushes

2. OPTIMUM OUTER-PERIMETER SECURITY BARRIERS

This diagram shows the outer block wall barrier, the hedgerow barriers, and the flowerbed barriers. Both the hedgerow and flowerbed barriers are nothing more than psychological barriers, but they serve the purpose and add a pleasant exterior landscape. The block wall is the only real physical barrier, but depending on the height and thickness of the wall, it too may only be a psychological barrier. Notice the "Escape Route" on the backside of the property. In this diagram, the wall has been deliberately weakened to allow a car to drive through it in an emergency escape without doing serious damage to the vehicle.

Figure 10

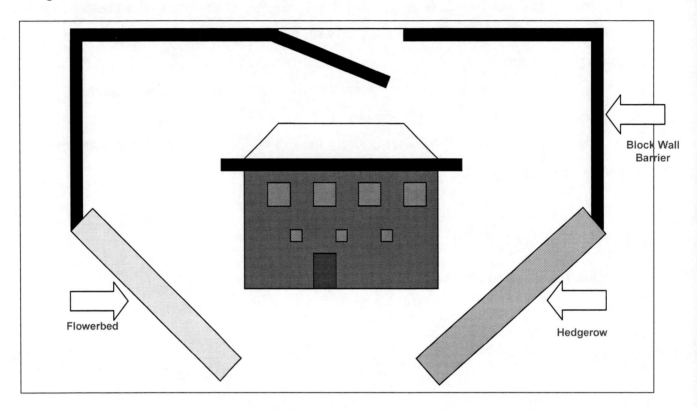

b. INNER PERIMETER

1. Primary door locks: Secondary door locks: deadbolts, window bars, barrier plants.
2. Use a passive infrared detection device and/or magnetic detectors that note the opening and/or closing of the door/window.

This Diagram displays a "Visual Concept" of a perimeter sector defense plan. This is how you must think when you consider your Organizations Continuity Management and Security Programs."

Figure 11.

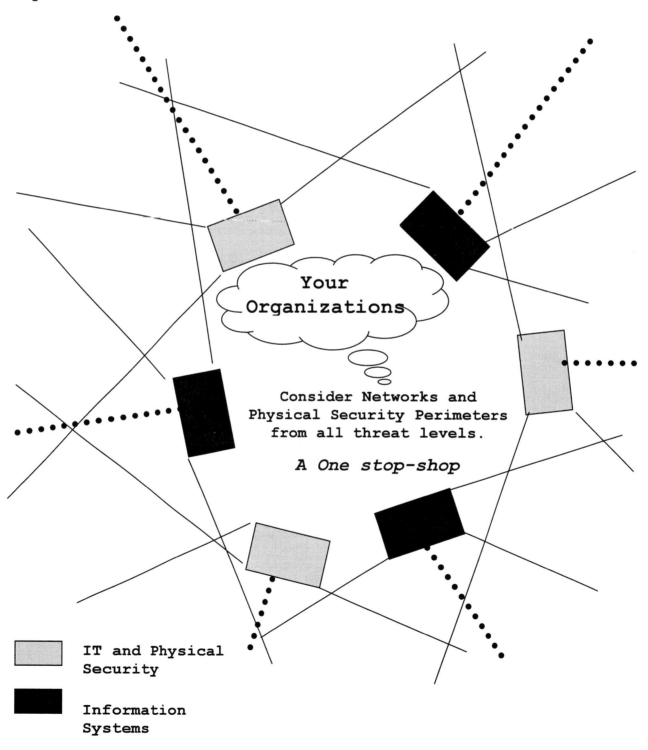

G. SECURITY LIGHTING

Emergency lighting is not just about having a reliable set of back-up lights. If you are in an emergency situation, you want lights that can continue to function in whatever unlikely situation may arise. Light systems should meet all your requirements for a back-up emergency lighting system and be entirely independent of your main electrical system, bright, water proof and intrinsically safe.

a. Security lighting is an integral part of any security system. It serves to stop intruders and to illuminate the area for observation.

b. Security lights need to be sufficient for illumination of the entire outer perimeter and for the interior of all buildings.

c. In areas where electricity may be intermittent, a back-up generator is a necessity.

d. Security light installations or upgrades should include: OFFICE SECURITY MANAGEMENT. Good office security procedures are paramount to securing property and personnel.

H. Visitors:

a. All visitors will be accompanied when on the premises.

b. Unknown visitors should be required to leave a driver's license or other form of pictured identification with the front desk.

c. Workmen without proper identification and authorization should not be allowed on the premises.

d. Janitorial and maintenance work in key areas must be constantly supervised.

e. Visitors and staff should have limited or controlled access to the manager's office.

f. All unknown visitors will be challenged in a low profile manner to determine the reason they are on the premises.

g. Visitors will not, under any circumstances, be permitted in cash rooms or areas where sensitive information is in residence.

I. Cash Procedures:

1. All cash must be kept in a safe.
2. Only two persons in the office are allowed to know the safe combination.
3. Under no circumstances will cash be kept outside the safe.
4. Cash is to be removed only for transfer or transactions.
5. Routes to and from locations requiring cash transactions will be varied.
NEVER USE THE SAME ROUTE TWICE.
6. Cash will be counted before it leaves the building and not produced until the time of transaction. Then and only then will it be removed from its hiding place.
7. Cash in transit will be concealed on your person, under clothing or in a briefcase with a carry strap, which can be affixed across the courier's chest. Doors and windows will be kept locked at any time the building is unoccupied. Two staff members will be required to check windows and doors prior to the close of work each day.

Appendixes

Appendix A

Example of a
Designation of Emergency Workforce Position for your COOP Program

---

TO: <EMERGENCY WORKFORCE PERSONNEL>

SUBJECT: Designation of Emergency Workforce Position

FROM: Deputy Director

1. This is to notify you that your position, <TITLE, SERIES, GRADE, PD#> has been designated as an Emergency Workforce Position in support of the XXX Headquarters, Continuity of Operations (COOP) Plan. For the purpose of the COOP plan, emergency workforce positions are those positions that must be occupied during a national emergency or mobilization without seriously impairing the capability of XXX to function effectively.

2. In this emergency workforce position, it is required that you perform assigned duties to support mission requirements during periods of national emergency or mobilization until relieved by proper authority. You are also required to be ready to perform these duties independently or in conjunction with other personnel in the same locality or at any assigned site within CONUS as designated by XXX management.

3. You should be aware that when notified of an emergency, you must attempt to report for duty, or if not possible, to contact <SUPERVISOR> to obtain reporting instructions. Further, if the alternate emergency workforce personnel are not available to report for duty, it should be assumed that all approved leave is automatically cancelled at such time and those employees who are on leave or are scheduled to take leave will be expected to report for duty.

4. Employees who occupy emergency workforce positions may be required to work in shifts of a minimum of 12 hours, in conditions which may include, but are not limited to 1-30 days of confinement, in

facilities which are not Americans Your Organization disabilities Act (ADA) compliant, and areas with limited lighting and no windows.

5. If you have a medical condition that prohibits you from occupying this position with or without reasonable accommodation, it is recommended that you immediately notify your supervisor. A determination will be made regarding the Agency's ability to provide reasonable accommodation for this position in the event of an emergency. If the agency is unable to provide reasonable accommodation, you will be reassigned to a position that is not designated "emergency workforce."

6. If you do not agree to occupy this "emergency workforce" position, you will be reassigned to a position that is not designated "emergency workforce."

7. Emergency workforce personnel, who agree to occupy this "emergency workforce" position, but subsequently fail to honor the terms of this agreement, may be separated from Federal service under appropriate procedures.

8. If you believe that your position has been incorrectly designated as an "emergency workforce" position, you may submit an administrative appeal within 15 calendar days of receipt of this notice in accordance with the Instruction XXX-XXX, Chapter XXX. The Agency decision regarding any appeal is final and is not subject to review.

9. If you are a Ready Reservist and you agree to occupy this "emergency workforce" position, you will be transferred to the Standby Reserve or the Retired Reserve, or shall be discharged, as appropriate in accordance with Directive, Number XXX dated May XX, XXXX and Personnel Management.

10. Please sign and date the receipt acknowledgement memorandum attached to this notice and return to <SUPERVISOR>, no later than <DATE>. If you have any questions concerning this matter, please contact the COOP Officer, at, XXX-XXX-XXXX.

11. This position requires Emergency Workforce personnel. An Emergency Workforce is an incumbent of a position who must report for duty and can be deployed to ensure that the functions of the Agency continue without regard to length of time or duty location. The incumbent also may be required to take part in contingency exercises.

<SIGNATURE AND TITLE>

Acknowledge receipt   _____   _____
                              (Employee Signature)   (Date)

TO: <EMERGENCY WORKFORCE PERSONNEL>

SUBJECT: Designation of Emergency workforce Position

FROM: Deputy Director

I acknowledge receipt of the designation letter and have read its contents. Any questions that I had have been satisfactorily answered. I understand that I may be called at anytime to perform these duties. I understand that upon acceptance of this position refusal to perform these duties will result in initiation of disciplinary action up to and including removal.

_____   I agree to be assigned to this emergency workforce position, <TITLE, SERIES, GRADE>, and to perform the assigned duties in support of the XXX Continuity of Operations (COOP) Plan.

_____   I decline assignment to this emergency workforce position, <TITLE, SERIES, GRADE>, and I request that I be reassigned to a position that is not designated "emergency workforce."

_____   _____
     (Signature)   (Date)

Appendix B

Example of an

OCCUPANT EMERGENCY ACTION PLAN

PURPOSE is to provide Company Personnel as well as any Guests/Visitors with a plan that will facilitate safe and orderly evacuation in event of fire or any other emergency situation that could be detrimental to their safety and health.

GENERAL:

The guidance contained in this Plan is general in nature and is designed to develop a state of readiness in the event of an emergency condition. It is imperative that each Employee /Occupant know exactly what to do, when to do it, and where to go in the event of an emergency.

No plan can cover every conceivable contingency. Employees/Occupants must become familiar with the general provisions of this Plan and be able to apply them with common sense under the supervision/direction of Monitor Personnel. The basic premise which must be observed is that, in any emergency, the safety of the individual is paramount. Each Employee/Occupant must remain calm, follow instructions in an orderly manner, evacuate as directed, and assemble in designated areas.

Fire Prevention is an inherent and important part of everyone's job and requires alertness and cooperation from all individuals and agencies in the building. Bomb Threats, unlike Fire Hazards, are not subject to employee control; the presence of an actual explosive device or a warning may be real or fictional and the warning may be anonymous or from an identifiable source. Specific instructions for Fire Emergencies and Bomb Threats are also addressed in this Plan.

This Plan has been developed for the protection of Personnel, and Guest/Visitors to the building. The Plan will work efficiently only through strict adherence to the instructions and procedures prescribed herein and through an application of common sense.

It is required that Company Personnel be assigned to Key Command and/or Monitor Positions to effectively manage an Emergency Building Evacuation as indicated in this Plan. Each floor in the building has a "Team of Monitors and Zone Wardens" working together to effectively manage evacuations. Occupants are to take direction from them during evacuations. The Orange Safety Vest they wear during such an emergency identifies all monitors.

Key Positions are:

Designated Official
Alternate Designated Official Occupant
Emergency Coordinator
Alternate Occupant Emergency Coordinator
Floor Monitors
Alternate Floor Monitors
Zone Wardens, Alternate Zone Wardens
Stairwell Monitors, Alternate Stairwell Monitors
Impaired Persons Aide, Alternate Impaired Persons Aide
Medical Coordinator
Security Personnel

A description of their Positions and Responsibilities is provided in this Plan.

IMPORTANT: It may become necessary for a Zone Warden to assume the duties of a Floor Monitor; therefore, they should be prepared to assume the role and responsibility of a Monitor Position that appears to be unattended. Designated Primary and Alternates are required for each position.

ALARM ACTIVATION:

The decision to activate the Plan, which may involve the evacuation of the building, is made by the Designated Official or his/her Designated Representative compound on the best information available and the advice of Local, State and/or Company Officials. When there is an immediate danger to Occupants, such as fire, the signal to evacuate the Building will be by activation of the Fire Alarm System. Any individual with knowledge of an emergency will then report the circumstances to the Occupant Emergency Coordinator who will activate the Fire Alarm.

DEFINITIONS AND RESPONSIBILITIES:

Command Center: In the event of a Building Evacuation, the Designated Official will establish a Command Center in the parking area. The Command Center may relocate to an alternate position determined by Security. Dependent on the type of emergency, a Command Center may be established in the Lobby Security desk. Noise from the fire alarm may prevent this option. If an evacuation is necessary in other buildings in the Compound, they will take instructions from the Security or his representative.

Floor Monitors will report to and assemble at the Command Center after securing, ensuring evacuation of their floors, and securing an accurate count. The accountability report may be radioed to the Command Center or given in person. Keep all radio transmissions short and to the point to avoid tying up the radio net. They will stand by to receive any available information about returning to the Building, or further action to be taken, etc.

If the emergency does not require Building Evacuation, noise permitting, the Command Center may be established in the Lobby.

Designated Official (or Alternate) in your Organization/Agency has the:

Responsibilities

- Supervision of the Occupant Emergency Coordinator. Overall responsibility for all planning and execution in connection with the Occupant Emergency Plan.

Facilities and Engineering - Building Manager:

Responsibilities

The Real Estate, Facilities, and Supply Services, will be responsible for the Company's building/compound. *********

Occupant Emergency Coordinator (or Alternate):

Definition - The Occupant Emergency Coordinator is responsible to the Designated Official for the efficient functioning of this Plan within *the Compound.* An individual designated by the Designated Official who acts for and in the absence of the Designated Official.

Responsibilities -

Assist the Designated Official in the planning and execution of Plans.
Acts as the Agency Liaison with the WHS Building Manager as required, for the planning of and execution of this Plan.
Assesses possible emergency situations and formulates plans for dealing with them. Identifies Occupant movement routes and establishes movement procedures to effect the evacuation of the Building. During drills and actual emergencies, Supervises and directs movement of all Personnel within, into, and/or out of the Building as necessary.
In cooperation with local Representatives, schedules and/or conducts Evacuation Drills. Schedules Training Exercises for Tenant Elements in the Building. Assures Building Security procedures are activated when evacuations occur. Assist Building Security when evacuations occur.

During emergencies, reports to the Command Center. Is the Command Center Monitor and receives reports from all Primary Floor Monitors as to the disposition of their floors.

Supervises and provides direction to all Monitors. Assigns all Monitors and related Personnel as required. Ensures that vacancies in Monitor Positions are filled as they occur.

Assures that all Building Occupants have been provided with Building Evacuation Procedures, informing them of procedures to follow during emergencies and that the Occupants comply with the Procedures as indicated.

Maintains a current listing of Occupants who, due to physical constraints, prevents them from evacuating the Building, will be reporting to a Safe Haven Room.

Floor Monitor:

Definition - A designated individual on each Floor of the Building, reporting directly to the Occupant Emergency Coordinator.

Responsibilities -

Responsible to the Occupant Emergency Coordinator. As directed assist with the planning and execution of this Plan.

Provides Occupants on his/her Floor with Building Evacuation Procedures, routes to follow during Evacuations, and procedures to follow during emergencies.

Supervises and provides direction to all Monitors on his/her Floor. Assigns all Monitors and related Personnel as required. Ensures that vacancies in Monitor Positions are filled as they occur.

With the assistance of additional Monitors assigned, directs the evacuation of all areas on his/her Floor by preserving order and discipline during actual emergencies, drills and exercises.

Identifies Occupants on his/her Floor that, due to physical constraints preventing them from evacuating the Building, will be reporting to a Safe Haven Room, and require assistance during Evacuations. Responsible for maintaining a current listing of these individuals and provide the list to the Occupant Emergency Coordinator.

Reports to the Occupant Emergency Coordinator at the Command Center, the disposition of his/her Floors during Emergency evacuations, drills, and exercises.

Periodically inspects EXIT signs and evacuation routes.

Alternate Floor Monitor:

Definition - Person designated on each Floor of the Building, who acts for and in the absence of the Floor Monitor, reporting directly to the Occupant Emergency Coordinator.

Responsibilities -

In the absence of the Floor Monitor, reports to the Occupant Emergency Coordinator the disposition of his/her Floors during emergency evacuations, drills, and exercises.

As required/necessary, assists the Floor Monitor in the execution of his/her responsibilities.

Zone Warden: (or Alternate)

Definition - Person designated to assist in and assure the evacuation of the Floor Zone assigned.

Responsibilities -

Verifies that their Zone has been evacuated. Reports to Floor Monitor the disposition of his/her Zone, assists with other tasks as necessary, and/or departs the Building.

If a Stairwell is located within the Zone assigned, checks the condition of the Stairwell, provides assistance to the Stairwell Monitor/Occupants as necessary, redirecting Occupants at the overcrowded Stairwell to the next available unobstructed Stairwell, and notifies the Floor Monitor of the disposition of the Stairwell.

If a Safe Haven Room is located within the Zone assigned, verifies that Occupants have reported there, provides assistance to the Room Monitor/Occupants as necessary, and notifies the Floor Monitor of the disposition of the Room/Occupants.

Stairwell Monitor: (or Alternate)

Definition - Person designated to assist in evacuation of the Occupants at the Stairwell.

Responsibilities -

Reports to the assigned Stairwell and directs the orderly flow of Occupants entering the Stairwell. If Stairwell becomes overcrowded, redirects Occupants to the next available unobstructed Stairwell if available.

Closes the Stairwell Door, notifies the Zone/Floor Monitor of the disposition of the Stairwell, assists with other task as necessary, and/or departs the Building.

Impaired Persons Aide: (or Alternate)

Definition - Person designated to assist Occupants who report to the Safe Haven Room.

Responsibilities -

Reports to the assigned Safe Haven Room to assist Occupants who physically cannot evacuate the Building via a Stairwell, due to a permanent or temporary physical condition, or have difficulty walking distances.

Impaired Persons Aides are to remain with Occupants that report to these Rooms until an ALL CLEAR condition is sounded, or evacuation is required. If evacuation is required, Emergency Services Personnel will evacuate the Occupants needing assistance. The Impaired Persons Aides are to take direction from Emergency Services Personnel and provide assistance as necessary. Impaired Persons Aides are not responsible for evacuating the occupants.

Notifies the Zone/Floor Monitor who has reported to the Safe Haven Room, provides assistance to those Occupants as necessary, and notifies the Zone/Floor Monitors of the disposition of the Room/Occupants. Assists with other tasks as necessary and/or departs the Building.

Medical Coordinator: (on site)

Definition - Registered Nurse, under direct supervision of the Occupant Emergency Coordinator.

Responsibilities -

Provides Medical assistance, advises as necessary, and insures that appropriate assistance has been summoned. Renders medical care as required until Emergency Services Personnel arrive. Works with Emergency Services Personnel, responding as necessary.

Provides report of Medical Services provided to the Occupant Emergency Coordinator.

Security Personnel:

Definition - Personnel assigned to the Security Division.

Responsibilities -

Security Personnel at assigned positions maintain control of ingress/egress to and from the Building at all Entrances and Exits.

Direct Visitors and/or Guests out of the Building.

Allow Emergency Services Personnel to enter the Building, but prohibit entry to any other unauthorized individual into the Building.

Upon receipt of an "ALL CLEAR," allow re-entry and as appropriate, check Photo Identification Badges.

Advise visitors to familiarize themselves with evacuation procedures.

Building Occupants:

Definition - Federal Civilian Employees, Military Personnel or Contractors whose place of duty is in any building in the Headquarters Compound, any Guest or Visitor.

Responsibilities -

Know the evacuation routes, and evacuate in a calm and orderly manner, obeying all instructions given by the Monitors.

If possible, secure work area prior to evacuating. Comply with Activity and/or YOUR ORGANIZATION Security Requirements for securing Classified/Security items. Know the locations of Fire Alarm Pull Stations. Occupants are prohibited from removing vehicles from the Parking Areas.

Upon activation of the Fire Alarm System, an Audible and Visual warning is given to alert Occupants to Evacuate the Building.

Elevators will NOT be used in Evacuations

Illuminated Red-and-White EXIT Signs identify EXITS. All Personnel will Exit the Building using their assigned Evacuation Routes (posted).

All Occupants are to Evacuate *(NO EXCEPTIONS)* and EXIT the Building. No Occupant in the Building is exempt from Evacuating. Individuals who do not evacuate are in Violation of the Law. Managers and Monitors are to encourage and enforce Occupants to move swiftly, avoid strolling, and carrying on conversations during an Evacuation. All efforts are to be directed toward getting out of the Building.

Occupants are to evacuate those Floors via the closest Stairwell, which leads to the ground Floor, or if exiting the compound of any building, use the closest exit. EXIT the Building using routes and exits that you have been briefed on.

Personnel are NOT to enter or assemble in front of other nearby Buildings. This raises Security concerns for those Buildings as well as blocks access for those who have business in those Buildings and blocks the EMERGENCY EXIT from those Buildings.

Do not attempt to travel between Floors if you are not on your Floor when the Alarm System is activated - EVACUATE THE BUILDING!

ALARM AND ACTION:

Upon activation of the Fire Alarm System, the Alarm System will activate on all Floors, in the Stairways, and the Elevators.

All Monitors will report to their Assigned Positions.

The following Personnel report to the Command Center:
Designated Official/Alternate
Occupant Emergency Coordinator/Alternate
Command Center Monitor/Alternate

Only the on scene Officer of the County Fire Department authorizes/grants re-entry to the Designated Official and/or the Occupant Emergency Coordinator. The Designated Official and or the Occupant Emergency Coordinator will then announce an "ALL CLEAR."

At no time prior to re-entry authorization will Occupants be allowed to enter the Building for any reason. Although this ruling may be an inconvenience, during an Emergency the Safety of the individual is paramount.

IMPORTANT: Each person assigned as a Monitor, including any Occupant not assigned to a specific responsibility, should be prepared to assume any responsibility that appears to be unattended.

ASSEMBLY AREA(S):

Know the Assembly Area(s). Report to the Assigned Assembly Area for accountability; remain in the Assembly Area(s) until otherwise advised. Occupants Evacuating the Building are to proceed to their Floors Assigned Assembly Area, which are identified as follows:

Assembly Areas are assigned as follows:

Bldg 25	Zone 5
Bldg 26	Zone 4
Bldg 27	Zone 3
Bldg 28	Zone 2
Bldg 29	Zone 1

Upon arrival to the Assigned Assembly Area, all Personnel are to report to their Supervisor for accountability. Any individual who has not vacated the Building and/or reported to the Assembly Area, is to be reported as unaccountable by the Supervisor to the Floor Monitor, who will notify the OEC, advising who the individual is, and that the individual is possibly still in the Building, and their possible (last known) location. All personnel <u>will remain</u> in the Assembly Area(s) until they receive further information OR the ALL CLEAR is given to allow re-entry to the Building. <u>All personnel must be accounted for.</u>

Appendix C

Example of Receiving and Reporting BOMB Threats:

An individual receiving a Bomb Threat by telephone must remain calm, keeping the caller on the line as long as possible.

If the caller does not volunteer the location and possible detonation time of the bomb, he/she must be asked for this information. At the same time as receiving the call, try to get another Employee to notify the Security Guard Control Center (XXX-XXX-XXXX) of the incident. The Security Guard Control Center will notify the following: Force Protection Agency XXX-XXX-XXXX or County Police 911 and the Designated Official XXX-XXX-XXXX.

Using the Bomb Threat Report, the individual will write down the message given by the caller, word-for-word if possible, listen closely to the voice for quality, accent or speech impediment, and identify if the caller is male or female.

During the bomb threat, the authority for Evacuation is the responsibility of the Designated Official and/or the Occupant Emergency Coordinator. Room Occupants can normally identify foreign or suspicious objects, which do not appear to belong in the Area. Such items <u>should not</u> be touched, moved, or disturbed in any manner. Floor Monitors will report the location of suspicious items to the *Force Protection Agency at XXX-XXX-XXXX* and the Occupant Emergency Coordinator at the Command Center (location, as indicated above, location depends on circumstances and whether or not Evacuation has taken place). The Occupant Emergency Coordinator will pass any information he/she receives from Floor Monitors onto the Designated Official, Command Center Monitor, Police, and Fire Departments.

If partial or full evacuation is required, messenger may give via the telephone or notification to the Floor Monitors.

BOMB THREAT PROCEDURES

Bomb threats may be received in several ways -- by telephone, by mail, and by messenger.

If you receive a bomb threat by telephone:

Keep the caller on the line as long as possible; talk to the caller; ask him/her to repeat the message.

Try to find out:

WHEN .... when will it go off?

WHERE .... where is it, where do we look?

WHAT................................... what does it look like?

WHY...................................... why are you doing this?

WHO...................................... who are you, what group do you belong to?

WHERE................................. where are you?

Record the time and date of the call.

Pay particular attention to background noises such as motors running, music, laughter, etc.

Listen closely to the voice (male, female), voice quality (calm, excited), accents, and speech impediments.

Report the call immediately to the *Force Protection Agency* at XXX-XXX-XXX, *Security Guard* at XXX-XXX-XXXX and the *Designated Official* at XXX-XXXX and/or the *Occupant Emergency Coordinator* at XXX-XXX-XXXX.

EVACUATION/BOMB SEARCH CHART

Building: _____ Date: _____

Floor/Area	Time Evacuated	Searched	Remarks

Note: This chart provides a useful tool for reviewing the effectiveness of an evacuation and/or search. Remarks may include who searched the floor or area, where people were relocated, and any unusual circumstances encountered.

BOMB THREAT REPORT
Exact Wording of Threat

Questions To Ask:

1. When is the bomb going to explode?
_____
2. Where is it now?
_____
3. What does it look like?
_____
4. What kind of bomb is it?
_____
5. What will cause it to explode?
_____
6. Did you place the bomb?
_____
7. Why did you place the bomb?
_____
8. What is your address? Where are you?
_____
9. What is your name?
_____

* If voice is familiar, whom did it sound like?

| Sex of caller: | Age: | Race: | Length of call: | Date of call: |

Did the caller appear familiar with the office or building by his description of the bomb location?

| | Yes | | No |

CALLER'S VOICE:

☐	Calm	☐	Laughing	☐	Lisp	☐	Disguised
☐	Angry	☐	Crying	☐	Raspy	☐	Accent
☐	Excited	☐	Normal	☐	Deep	☐	Stutter
☐	Slow	☐	Distinct	☐	Ragged	☐	Rapid
☐	Slurred	☐	Nasal	☐	Loud	☐	Soft
☐	Clearing	☐	Cracking	☐	Deep	☐	Familiar*

BACKGROUND SOUNDS:

☐ Street Noises
☐ Airport Noises

☐ House Noises
☐ Motor
☐ Office Machinery
☐ Factory

☐ PA System
☐ Long Distance
☐ Clear
☐ Local

☐ Animals
☐ Music
☐ Other (specify) _____

☐ Well Spoken
☐ Foul
☐ Irrational
☐ Incoherent
☐ Message Read by Threat maker

Remarks

| |
| |

Key Factors

Name	Position	Phone No.	Date

KNOW THE PLAN

Read, review, rehearse, and exercise the plan

Know your responsibilities
Know the location of the nearest fire alarms, emergency exits, and Safe Havens and Assembly Areas
Know who is designated the Building Emergency Coordinator
Preparation and participation are the keys to your safety and survival
There is no *"I"* in the word *"TEAM"* ... we all need to look out for one another

EMERGENCY SITUATIONS

Bomb Threat *
Suspicious Package *
Fire *
Chemical, Nuclear, Biological, Radiological (CBRN's) Emergency */**
Tornado or other Severe Weather **
Police Emergency **
Area/Installation Evacuation
Workplace Violence
Utility Failure (Power, Gas leak)
Terrorist Threat/Hostage Incident

Appendix D

Example of a SHELTERING IN PLACE:

    a. The Federal Government has determined that "Sheltering in Place" (SIP) is the most prudent way to ensure employee safety. SIP is a method to ensure employees remain in a sheltered location until official word is received from government or local authorities that it is safe to depart a building or other location that has served as a safe haven during an area emergency.

    b. <u>SIP (Calm)</u> This category encourages employees to remain in place until any <u>temporary area emergency</u> has been resolved. These emergencies <u>may</u> include such events as a local hazardous fuel tank vehicle mishap, or any hazardous condition that could affect air quality or the ability of traffic to move safely along major highways.

    c. <u>SIP (Threat)</u> This category requires employees to <u>indefinitely remain in the occupied sheltered place</u> until official word is received from management in your organization, local authorities, or local television and radio affiliates that the danger has passed. <u>At no time should employees attempt to leave the sheltered location under this threat until officially informed to do so.</u>
    Employees will be given specific instructions as to options available to them and when employee release may be possible. SIP is a serious matter. Employees are encouraged to have <u>preplanned arrangements</u> in place for both the office and home such that office and family concerns during any SIP event will be minimized.

*** SHELTER IN PLACE*

HOW YOU WILL BE NOTIFIED

Building Alarm System/ PINS
Supervisor or Building Emergency Coordinator
Telephone
Police/Fire Emergency vehicle loudspeaker
Personal observation of an event

IMMEDIATE ACTIONS AND FIRE EMERGENCIES

Remain Calm/Stay Alert
Activate fire alarm, notify the fire Dept (911) and notify Security
Secure your work areas following Activity/Security Procedures for securing items and material
Close all doors. Do not lock them
Evacuate the building through designated Emergency Exits. <u>Do Not Use The Elevators</u>! The situation will determine its use. Report to your appropriate assembly area for roll call
Account /Assist all impaired persons to Safe Haven areas
Remain in your assembly area until otherwise directed by proper authorities

TORNADO OR OTHER SEVERE WEATHER CONDITIONS

Remain calm

Move away from windows and doors. Move toward the center core of the building. If possible, seek shelter in lower areas of the building

Do Not Use the Elevators!

Monitor the local weather forecast

Follow directions of Security Personnel and Management

Call the office Hot Line at XXX-XXX-XXXX

The Designated Building Official will inform personnel of any early release

WORKPLACE VIOLENCE

Where possible, maintain a sense of calm with your fellow employees

Report erratic behavior of co-workers to your supervisor

Do not engage in one-on-one contact with an employee who is visibly distraught or unnerved

If possible, obtain the name of the employee. Notify Security for further guidance

Attempt to observe the employee for as long as you can without endangering your own safety

Depart the area immediately if told to do so

Hostage/Barricade Situation

UTILITY FAILURE

Power failure

Remain in the office and open blinds (if necessary for additional light)
Turn off all computer and related equipment
Notify supervisor of incident*
Ensure that security personnel have been notified
Await further guidance from senior officials
Should a utility failure occur during non-working hours, access to buildings will be denied to all persons unless approved by security personnel

Gas leak

Call Security
Notify supervisor of incident
Evacuate building to your designated assembly/rally point

*Note: Most phones will not work if power is lost.

Sheltering on Place
(Immediate or Imminent Threat)

Employees must immediately move away from all doors, windows, and ventilation systems that circulate outside air into the building.

Used when a dangerous threat is imminent could result in sickness, injury, or loss of life.

Civil Emergency "First Responders" will provide directions based on the nature of the emergency to ensure all steps are taken to avoid loss of life or serious injury by way of airborne contaminates, blast, or heat effects of an unknown, known, or undetermined threat or threat source.

Sheltering in Place
(Calm/Non-Life Threatening)

This sheltering category reduces the amount of vehicular traffic on area roads. This situation is normally used to allow local emergency vehicles to rush equipment and personnel to the site where an incident has occurred.

The situation causing the Sheltering in Place order has not been deemed so dangerous as to be harmful if employees remain in place.

Employees are free to move about *inside* the building to conduct normal business.

FORCE PROTECTION CONDITIONS (FPCON's)

FPCON Definitions: NORMAL = ROUTINE OPERATIONS

NORMAL   - No general threat
ALPHA   - General threat, nothing specific
BRAVO   - Increased threat, maintained several weeks
    - Some CHARLIE measures
CHARLIE   - An incident has occurred or a threat is imminent
        - Maintained for short periods
            - Some DELTA measures
DELTA   - An attack has occurred and others are likely
    - Specific locations may be affected

BASIC PLAN REMINDERS

Comply with open/lockup procedures. Last person out secures the building! End-of-day security check forms will be maintained & posted outside each office.

All personnel will comply with FPCON measures & actions.

During higher FPCON's, access to all buildings will be limited, monitored, and may be denied. However, emergency exits will be available for evacuations if necessary. Everyone entering the Buildings will be checked for and must have a valid picture ID, every time, regardless if previously cleared or known by Security personnel or others present at the point of entry.

All personnel and visitors will cooperate with these procedures and display all required identification.

All ID badges must be worn above the waist.

AREA/BUILDING EVACUATION

Know if you are designated as a key and/or "Essential Person"
If not, remain in place. You will be advised to evacuate elsewhere, if safe to do so
If yes, know your individual organizational procedures for continued operations
If directed to evacuate the building for an extended period of time, ensure all classified/sensitive papers or equipment is properly secured

BASIC PLAN REMINDERS
(cont)
All personnel:

> <u>Know this plan</u>
> Maintain alert rosters-Emergency Telephone Call Lists
> Be aware of and report anything suspicious
> Know the Assembly Areas
> Account for all personnel and report same to On-Scene authorities
> Assist others; consider impaired personnel, as well as visitors

Emergency instructions from Civil Authorities must also be followed to ensure your safety

FIRE EMERGENCY

Activate nearest Fire Alarm Pull Station, or notify the Fire Dept (911).
Notify Security, retrieve your Escape Hood, and take it with you.
Secure your work areas following established Activity/YOUR ORGANIZATION Security Procedures for securing classified items & material.
Close all doors, *do not lock them.*
Evacuate the Building through designated *Emergency Exits* and report to your assigned assembly area for roll call.
Assist all impaired persons to known Safe Haven areas if requested.
Do Not Use the Elevators!

EVACUATION ASSEMBLY POINTS

Appendix E

A Checks List for Your Business Recovery Plan

The first step of your Business Recovery Plan (BRP) is senior management's approval and authorization. Recovery activities (cost) are these requirement budgets.

Let's start from the top. Is there a BRP; furthermore, is the BRP written and/or documented?

That's not a YES or NO question.

Physically check and review to see if your organization has a BRP or take notes as you read the checklist.

The next question is has the plan been updated within the last year? Yes or No, there is no middle ground here!

With the plan in your hand, ask yourself a question, is someone assigned to manage the BRP?

Furthermore, are there facilities, personnel, and operational supports assigned to the plan?

Now we're getting into the good stuff.

Does the BRP include the following sections: Incident Management, Responsible company officer, Personnel responsible for updates, Response, Recovery, Restoration, Plan Exercise, Plan Maintenance, Business Recovery Teams, and Contact Information?

Management must fully understand the BRP equipment and personnel requirements.

Does the Business Recovery Plan identify hardware and software critical to recover the Business and/or Functions?

Does the Business Recovery Plan identify necessary support equipment (forms, spare parts, office equipment, etc.) to recover the Business and/or Functions?

If your facility is inoperative? This question is critical.

Does the Business Recovery Plan require an alternate site for recovery, i.e. Hot and/or Cold sites?

You must also consider external functions and support activities.
Does the Business Recovery Plan provide for mail service to be forwarded to the alternate facility?

Does the Business Recovery Plan provide for other vital support functions, i.e. alternate employee payroll systems?

Are all critical or important data required to support the business being backed up? Are they being stored in a protected location (offsite)?

After you finish reviewing your BRP, investigate this next question; be complete and accurate.

Is a walk-through exercise of your Plan done at least annually? This should include a full walk-through as well as "elements" of your plan (i.e. accounts payable, receivable, shipping and receiving, etc.

Does the walk-through element exercise have a prepared plan, which includes Description, Scope, and Objective?

Another important question. Is there a current copy of the Business Recovery Plan maintained off-site?

Do all users of the Business Recovery Plan have ready access to a current copy at all times?

Is there an audit trail of the changes made to the Business Recovery Plan?

Do all employees responsible for the execution of the BRP receive ongoing training by your organization's Emergency Management?

In the second step, your BRP is completed.

The organization's planners must assess and evaluate Recovery Effectiveness.

Senior Management must approve the Business Recovery Plan. Managers must maintain the master copy of the Business Recovery Plan and an audit trail of the changes made to a Business Recovery Plan.

Security is equally important for an organization's planning. During security activity, is protecting company assets and employees highlighted in your plan? Safety is always first.

Do all aspects of physical and logical security at the alternate sites conform to your current security procedures?

Is the physical and logical security at the alternate site at least as stringent as the security at your organization's main location?

Will the business activities continue?

Have all employees and their alternates responsible for executing a manual work-around for a mechanized process been identified in the Business Recovery Plan and properly trained?

Conducting a true assessment: has an independent observer documented the simulation exercise(s) noting all results, discrepancies, exposures, action items, and individual responsible, etc.?

Was a debriefing held within a reasonable period of time (typically two weeks) after the simulation exercise(s) to ensure all activities have been accurately recorded?

Evaluations findings: Did the exercise coordinator publish a simulation exercise(s) report within a reasonable period of time (typically three weeks) after the completion of the simulation exercise(s)?

Did the exercise report include what worked properly as well as any deficiencies and recommendations for improvement and the responsibility and due date for the development of the Corrective Action Plan?

Was a Corrective Action Plan developed by the Exercise Team to address any deficiencies identified by the exercise?

Is there a retention plan for the Exercise Plans and Corrective Action Plans (minimum retention 2 years)?

Has a walk-through element exercise been performed at least quarterly?

Did each walk-through element exercise have a prepared plan which includes Description, Scope, and Objective?

There are new applications that can support system issues when there is a change in hardware, software, or a process that might impact the Business Recovery Plan. Is the Business Recovery Plan reviewed, updated within 30 days of the changes, and Signed-Off by Officer by Whom, Name, and Date?

Compound on the Joint Assessment: Has the Team determined that the Business Recovery Plan is effective?

Has the component Business Recovery Plan been approved by the owner(s) of the Business Function(s)?

Has the entire Business Recovery Plan simulation exercise been performed at least annually?

Has the Corrective Action Plan been completed and closed?

Did the Business Recovery Plan simulation exercise have a prepared plan, which includes Description, Scope, and Objective?

Did the component Business Recovery Plan simulation exercise meet the acceptable Recovery Time Objective set by management?

Compound on the Joint Assessment: Has the Team determined that the Business Recovery Plan and Exercises have met all requirements to provide reasonable assurance that the plan will work?

Does the Business Recovery Plan specify the maximum acceptable Recovery Time Objective (RTO)? Plans and manages the relocation.

Does the Business Recovery Plan specify the Level of service (which the business owner has agreed to be acceptable) to be provided while in recovery mode?

Have all changes relating to RTO in the Business Recovery Plan been approved by the process owner?

Appendix F

An example (Business Continuity Plan)

International has critical business functions that must be executed or quickly and efficiently resumed during a Disaster. Although Disasters cannot be predicted, planning for the continuation of business during such conditions can mitigate the impact of the disaster on our employees, facilities, and the company's overall image.

Our Business Continuity Plan must be sound and each branch manager will prepare a site-specific continuity plan with mission essential functions that his branch must perform during a disaster. Those plans will be included in the International's Business Continuity Plan. These plans will be important resources for ensuring that we continue to provide essential service to our customers, employees, and suppliers.

Approvals

Approved:

Title_____

Signature_____ Date/Time_____

Approved:

Title_____

Signature_____ Date/Time_____

Approved:

Title_____

Signature_____ Date/Time_____

Approved:

Title_____

Signature_____ Date/Time_____

BASIC PLAN

I. PURPOSE

A. Overview. To provide an overview of your organization's policies and organization for responding to emergency situations in a coordinated and effective manner to protect the lives and property by:
Assigning responsibilities to divisions and individuals in our organization to implement and respond to emergency activities and/or routine problems in our organization.
Defining organizational relationships and lines of authority to explain how actions are to be coordinated.
Describing how businesses' emergency response organizations and other resources will protect people and property during emergency or disaster situations.
Identifying the available resources in your business for both the response to and recovery from emergency or disaster events.
Defining the steps necessary for mitigation efforts during response and recovery activities.

B. Emergency Management Coordinator has the primary responsibility for emergency management activities in your organization.

This Plan is corporate wide in its relevance and applicability to ensure that all available resources within the organization are brought to bear on any emergency or disaster. When the emergency or Disaster exceeds your organizational response capabilities and resources, assistance will be requested through the local Emergency Management agencies in accordance with the Emergency Operations Plan, the Federal Emergency Management Agency (FEMA), the Federal Response Plan (FRP), and the FEMA Regional Response Plan.

II. SITUATIONS AND ASSUMPTIONS

A. Required Plan. In mandating the creation of emergency service in our company, every department is required to provide input, test, and maintain a current Business Continuity Plan.
Senior Management is prepared to commit all available resources to respond to any emergency and have mutual aid agreements among themselves and with adjoining jurisdictions should the need exceed the capability of any one particular organization or jurisdiction.
The historical causes of natural disaster or emergencies include floods, snowstorms, prolonged cold/freezing weather, drought, tornadoes, and hurricanes.
Manmade emergency or disaster situations include radiological incidents, transportation accidents, hazardous materials events, structure fires, explosions, power outages or shortages (electricity, natural gas, etc.), and the loss or shortage of other essential public services (potable water and sanitary sewer).
The Business Continuity Plan acts as a guide for the development of the more operationally oriented functional annexes, which follow it.
All organizational emergency response resources are considered available for responding to emergency events.

III. CONCEPT OF OPERATIONS

A. Integrated Emergency Management System. This all-hazards emergency operations plan utilizes an Integrated Emergency Management System (IEMS), which accounts for emergency management activities before, during, and after emergency operations. IEMS separates emergency management into the following four categories or "Levels" of emergency operations:

Mitigation: Activities designed to either prevent the occurrence of an emergency or long-term activity to minimize the potentially adverse effects of an emergency.

Preparedness: Activities, programs, and systems which exist prior to an emergency, are used to support and enhance response activities during an emergency or disaster.
Response: Activities and programs designed to address the immediate and short-term effects of the onset of a disaster, helping to reduce both casualties and property damage and to speed recovery.
Recovery: Activities that restore systems to normal. Short-term recovery actions assess damage and return vital life support systems back to minimum operating standards. Long-term recovery operations may continue for years after the disaster or emergency event.

B. Management Role. Senior Management has the primary responsibility for protecting the lives and properties of their organization from an emergency or disaster event. Organization Emergency Management are listed and described below.

Company Position	Emergency Position	Location

C. In emergency or disaster situations, which exceed the organization's capabilities, local and State Emergency Services will provide direct service, lead our efforts, and possibility act as a channel for obtaining and providing additional resources. We will continue to operate the EOC and conduct organizational command and control.

When local government agencies provide emergency assistance, which may include on-site representation, the overall command and control authority remains with the local jurisdiction, unless local control is relinquished or if State or Federal law requires the transfer of authority to a specified State or Federal agency.
All State departments and agencies with emergency responsibilities are to provide direct assistance to local jurisdictions where possible and to participate in local EOP activities.
We must immediately recognize the capacity of catastrophic disaster and evaluate the organizational response capabilities to enable a more pro-active response.

We must identify the organization's requirements, Mission Essential Functions (MEF), and priorities to limit the impact on our organization.
State and local governments will maintain direction and control over disaster response operations when the FRP is in effect.

D. Direction and Control - This is a critical emergency management function which enables emergency operations managers to analyze and direct efforts in a more effective and resourceful manner. This Plan utilizes a centralized management facility, or Emergency Operations Center (EOC), to facilitate policy and decision-making, coordination, and the overall direction and control of emergency responders. The organization EOC is the primary and centralized location for the direction and control of emergency response activities in your company.

Where is the Businesses EOC located? The address and directions.

The facility's communication's center is manned on a 24-hour, seven day per week basis. The facility has ample, additional space in the EOC to assemble emergency management staff and response personnel during an emergency or disaster.

    E. Warning and Alert - This emergency function enables your organization's emergency management organization to provide timely forecasts to the proper public officials and the public of all hazards which will require emergency response actions. It is essential that vital warning and alert information be made available to allow emergency responders and the public to take appropriate actions to avoid death or injury as well as damage to property.

What is your organization's Warning System, i.e. alarm, horns, etc.?
The Warning Point is staffed on a 24-hour, seven day per week basis.

III. ORGANIZATION AND ASSIGNMENT OF RESPONSIBILITIES

    A. The Organization's Emergency Structure has the following emergency management responsibilities:
Develop, distribute, and update local Emergency Occupancy Plan (EOP)s.
Ensure that all local EOPs within the company are coordinated with one another and that such plans comply with the West Virginia EOP.
Provide funding for emergency planning and management needs.

    B. Local Emergency Management. The Chief Executive Official of all political subdivisions (Mayors and President of the County Commission), with the appropriate authority of their respective governing body (City Councils and County Commissions) have the following primary responsibilities for local emergency management.

Assign and make available employees, property, and equipment for emergency operations of all levels.
Establish a local EOC and secondary control center or centers to serve as emergency command/control posts.
Enter into mutual aid agreements with other jurisdictions and organizations for the provision of emergency services.
Seek and accept services, equipment, supplies, materials, or funds by gift, grant, or loan for emergency management needs.

    C. State Of Emergency. During a declared local State of Emergency, these same officials have the following authority and responsibilities.
The activation of local plans and mutual aid agreements as is required to meet the needs of an emergency or disaster.
The implementation of provisions of local emergency readiness or resolutions as may be necessary.
The authorization for emergency workers to incur exposures in excess of the general protective action guidelines as defined by the United States Environmental Protection Agency (EPA).

    D. Coordinator/EMC. The Emergency Management Coordinator acts as the Executive Officer of the Organization and is responsible for directing its organization, administration, and operations. Additional authorities are as follows:

May develop mutual aid agreements with other agencies, or organizations for reciprocal aid when an emergency event exceeds the capabilities of the signatory parties.

Will develop and maintain the BCP coordinate and assist other divisions and agencies in developing EOPs or Suggested Operating Plans (SOP).
Will develop a rapid notification system to alert all emergency responders identified in this Plan.
Will maintain a public warning and notification system, which operates on a 24-hour basis.
Will develop a public information program to provide timely and accurate information to the public during all levels of managing an emergency event.
Will provide and coordinate appropriate training for emergency management and response personnel.
Will maintain accurate records of expenditures made by the organization OEM.
Will notify emergency management officials in other locations when there is a likelihood of an emergency in the company affecting their area.
Will notify the local emergency if assistance is required or when the situation requires additional resources.
Will review all known information about an emergency or disaster situation to determine an appropriate evacuation response.

Identifies emergency scenes where an Incident Commander (IC) may have already ordered an evacuation, delineates the perimeter, and verifies the degree of abandonment.
Locates assembly areas for evacuees lacking transportation.
Identifies and selects, in close coordination with law enforcement, evacuation routes from the risk area to designated mass care facilities.

  E. Communications Officer - This position is responsible for the management of all emergency communications systems. The Communications Officer will establish the protocols for all emergency communications operations:
Work with representatives of other emergency responders to develop communications procedures that will be responsive to organizational needs and be compatible with local procedures and equipment.
Identify and maintain a current listing of communications and warning resources within the organization, which are available to the EOC.
Identify private and public service agencies' resources (personnel, equipment, and facilities), which can augment the communications capabilities.
Develop or arrange training programs for all of the EOC's communications staff including repairing personnel.

  F. Public Information Officer (PIO) - The PIO advises the EMC of all matters relating to emergency public information (EPI) and must establish and maintain a working relationship with the local media. Other responsibilities include:

Prepares EPI packets for release.
Distributes pertinent materials to the local media prior to emergencies.
Ensures that the information needs of visually impaired, hearing impaired, and non-English speaking audiences are addressed.

  G. Mass Care Liaison (MCL) - The MCL reports to the EMC and upon arrival at the EOC performs the following responsibilities:

Assesses the emergency and recommends to the EMC on the number and locations of mass care facilities which should be opened.

Receives the list of available mass care facilities and notifies the appropriate facility contact of the possible need for services and facilities.
Selects mass care facilities for activation in accordance with EOP criteria.
When directed, coordinates the proper actions to ensure that all necessary mass care facilities are opened and staffed.

H. Health and Medical Coordinator (HMC) - The HMC is primarily called into the EOC during large scale emergencies or disaster events which could result in sufficient casualties and/or fatalities to overwhelm local medical, health, and mortuary service capabilities. When called to the EOC, the HMC accesses health and medical needs in a rapid fashion and:

Oversees and coordinates the activities of health and medical organizations and ensures that emergency medical teams at the disaster /emergency site establish a medical command post.
Coordinates with neighboring community health and medical organizations, as well as State and Federal officials, on assistance from other jurisdictions and levels of government.
Screens and coordinates with incoming groups, individual health and medical volunteers to ensure that positive identification and proof of licensure is made for all volunteers.

I. Administrative Support Coordinator (ASC) - The ASC reports to the EMC and will be one of the first emergency management staff members to be notified of an emergency and called to the EOC. The major responsibilities of the ASC during an emergency or disaster situation fall under four broad categories.
Determining needs compound upon past experiences combined with preliminary damage assessments, prioritizing needs, and follow-up and tracking.
Obtaining supplies and evaluating requests against known supply levels.
Maintaining financial and legal accountability of all resource transactions.
Distributing goods and services to sites where most needed and useful.

J. Tasked Organizations - These are additional organizations and agencies that support your organization, which as a result of their function or expertise will be tasked to provide emergency or disaster response services. These groups include all law enforcement agencies including specialized units such as fire service providers, EMS providers, and Hazardous Incident Response Team. To ensure their respective response capabilities, all tasked organizations are expected to:
Maintain their existing notification rosters, call-down lists, and SOPs to perform their assigned tasks.
Enter into mutual aid agreements, as may be appropriate.
Work with the EOC's Communications Officer to ensure that communications equipment and procedures are compatible.
Identify potential sources of additional equipment and supplies.
Provide for the continuity of their respective operations.

K. Other Organizations - There are additional organizations and agencies associated with your organization, which could provide critical assistance during particular emergencies or disaster events, which include:

American Red Cross (ARC) - The ARC has been designated as the primary agency for operating mass care facilities during disaster or emergency events, which result in the dislocation of people from their homes. Additional capabilities include:

Individual or mass feeding centers
Emergency shelter
Distribution of food, clothing, furniture, and household supplies
Capacity to recruit, train, organize, and direct volunteers in emergency or disaster relief assistance activities

Salvation Army - This organization can provide resources and assistance during emergencies and disaster events.

Public Utilities - Each utility has emergency management plans, which provide for a system of priorities for restoring essential services on an "essential need basis."

<u>Utility Company Services</u>

Power Company, Electricity, Verizon Telephone, Alpha-Brothers Telephone, Local Natural Gas

Media - The television stations, radio stations, and cable television providers providing emergency warning services or other emergency messages to the public. The BCP has been coordinated with the other Office of Emergency Management.
       a. Radio Stations
       b. Television Stations
       c. Cable Television

5. *Additional Groups* - Civic groups, industrial and business associations, and churches have a large pool of talent and resources, which could be of assistance in emergencies. These groups will be contacted and cultivated as a part of the emergency management process to establish informal understandings to use these resources and skills.

IV. ADMINISTRATION AND LOGISTICS

    A. Procurement. Emergency response activities and operations are primarily a responsibility of local government, and as a result, such activities will be funded within the budgets of each local agency or organization, which has an emergency response or management responsibility. During emergency or Disaster response and management operations, there is quite often the need for supplies, equipment, and services, which are not readily available in existing inventories or through normal procurement procedures. The situations under which extraordinary procurement procedures may be used are as follows:

Procurement Prior To a Declaration Emergency. Every effort will be made to meet emergency or disaster needs from local government resources. Local officials will be contacted regardless of normal business hours to assist in obtaining needed resources, which are not readily available in local government inventories. However, unless the proper local official specifically authorizes it, normal procurement procedures will apply to all transactions.

Procurement After a Declaration of Local State of Emergency. Those resources determined by competent authority required to save lives and protect welfare which cannot be obtained from "regular" sources may be requisitioned using procedures outlined beforehand by the County Commission.

PLAN DEVELOPMENT AND MAINTENANCE

    A. EOP. As stated earlier, your Organization's Emergency Management has the primary responsibility for emergency management activities in your company, which includes the development and maintenance of the BCP. The Coordinator of Emergency Operations has the overall responsibility for developing and maintaining the BCP and will review the Basic Plan and its Functional Annexes on a regular basis to ensure their integrity.

Basic Plan. It is the responsibility of the Coordinator to ensure that the BCP is capable of meeting the emergency management needs of the organization. The Coordinator will review and update the Basic Plan on an as-needed basis. Any changes or modifications to the Basic Plan or any of the Functional Annexes will be recorded in the section provided in the preface to the Basic Plan.

Functional Appendices. This BCP includes functional Appendices, which are attached to it. Each of the annexes includes a discussion of the parties, which are responsible for reviewing and updating that particular section of the BCP.

Appendices

- Telephone directories
- Authority delegation information
- Emergency and Organizational Contacts Lists
- Vendor List
- Alternate Facilities, listing and contact names
- Support elements contact list
- Relocation reporting roster
- Maintenance Plan, Budget

Appendix G

Example of an Alternate Site Guide

Table of Contents

1. Directions To The Site   112
2. Map To The Site 112
3. Arriving And In Processing   116
4. Security Procedures   117
5. Facility Emergency Procedures   120
6. Communications   125
7. Family Readiness   128
8. Medical And Dental Care   129
9. Lodging   130
10. Dining Facilities   132
11. Morale Welfare And Recreation   132
12. Departure, Out-Processing 133

NOTE: XXXX Emergency notification and contact number is the Security Office at 1-800-XXX-XXX. Physical Security Office: XXX-XXX-XXXX.

1. Directions to the Site

Primary Route: Directions to the Facility

From Upper Marlboro, MD -

2. Alternate Route 1 - From Kendridville) to main gate

3. Maps to the Site

From Upper Marlboro, MD:

From Baltimore:

On Site Directions to Building XXX

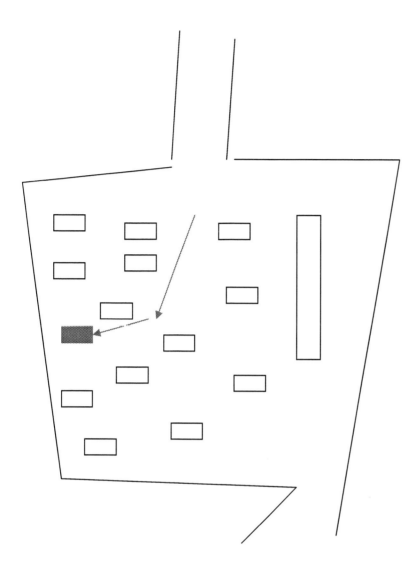

From Front Gate:

Straight in G Street, at second 4-way stop turn left. Building XXX is second building on the right. The entrance is on the East side of XXX. Use the entrance where the telephone is located (approximately in the center of the length of building).

Parking Areas and Entrances

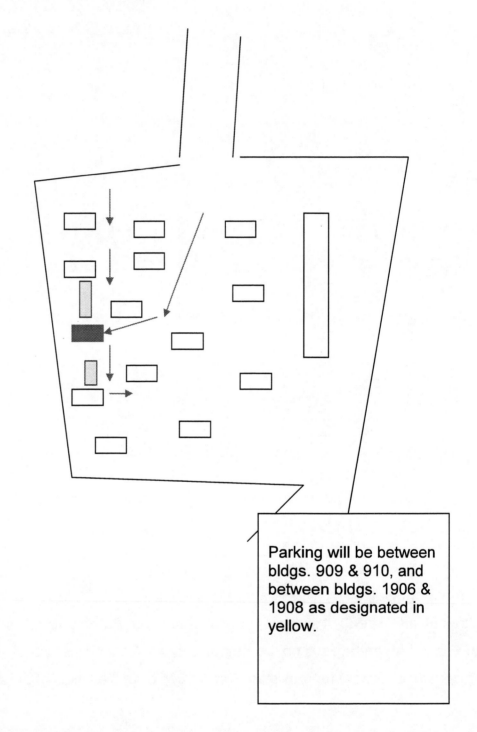

Parking will be between bldgs. 909 & 910, and between bldgs. 1906 & 1908 as designated in yellow.

EVACUATION PROCEDURES

1. The senior manager will make a real time assessment of the emergency. Recommended action(s) will be briefed to the Director or in their absence the Deputy Director. The decision to evacuate the building lies solely with the senior manager on site at the affected area.

2. Prior to any evacuation, each employee will be required to spot-check his or her area to determine if any hard copy classified information is unsecured. All attempts will be made to properly secure the classified information. However, at no time will personnel put the life or the life of their co-workers in jeopardy by trying to secure classified information. If necessary, classified information may be removed from the building. The classified information will remain in the custody of the person until it can be properly secured.

3. In the event of an evacuation, the elevator will not be used as a means to evacuate the building. The emergency stairwell located throughout the building will be the only authorized means to evacuate the building.

4. Upon an orderly evacuation of the building, personnel will move quickly to the south of Building XXX. The assembly point will be Pompey Drive across Bryce Avenue in the parking lot North of Building XXX. The senior manager will take a head count of personnel to ensure everyone has evacuated the building. The other assembly point will be in the middle of the parking lot at north end of Building XXX where the above process will take place. At no time will personnel be authorized to re-enter the building until the "All Clear" has been given.

5. Upon evacuation of the building, personnel will stay as far away from the affected area as possible. Personnel should not stand in the vicinity of windows and other areas that would cause injury to themselves from flying debris that is associated with a bomb blast, back drafts of a fire, and other disasters.

GENERAL EVACUATION PROCEDURES:

a. If safe to do so with minimal delay in exiting:

- Turn off electrical equipment.
- Place hazardous operations or materials into a safe standby mode.
- Close windows if applicable.
- Close door upon exiting the room, but <u>DO NOT LOCK</u> the door.

b. Do not carry coffee, food, soft drinks or items that if dropped could inhibit safe egress and cause slips, trips or falls.

c. Walk to the designated assembly point area by the designated route. If the route is blocked by unsafe conditions, take the nearest safe path out of the building.

d. Escort visitors, contractors, and vendors to the assembly point.

e. Do not smoke while you are leaving the building or at an assembly point.

f. Beware of and give the right-of-way to responding emergency vehicles and personnel.

g. Inform the senior manager of any hazardous situation that may have been encountered during the evacuation.

f. Each building occupant is responsible for being familiar with the evacuation route and the location of the assembly point.

g. The use of cell phones and other electronic equipment are not permitted during actual and/or drill evacuations.

h. Assist disabled or (physically, hearing, etc.) personnel during the evacuation of the building and stay with them at the assembly point.

i. Notify the senior manager of any missing personnel or suspected missing personnel.

j. Do not re-enter the building for any reason unless the "All Clear" has been given.

3. Arriving and In Processing

All Personnel Must: Have a Company badge for entry onto the Compound. Entry will be authorized using their Company's badge and sign in will be at the desk just inside the door. Once inside, a security area will be set up where you will show the required identification. You will then be issued a facility badge and will proceed to your designated work area. You will be taken or directed to a designated local motel for lodging and returned to the facility for duty.

Arrival by Provided Vehicle: Personnel will be required to show ID to gain access to the site. Follow the site map on page X to the facility and park in the authorized parking area. Enter the building through the side entrance, where the telephone is located. Once inside, a security area will be set up where you will show the required identification. You will then be issued a facility badge and will proceed to your designated work area.

Arrival by Bus: Some personnel may arrive at the site by bus directly to Site.

a. The bus carrying personnel arrives at the main gate. A Security Guard will board the bus. The on-bus escort officer provides the Police Officer with a manifest of all personnel on the bus including the driver, clearance information included. When directed you will enter the compound and proceed to Building XXX. If no manifest is provided then ALL personnel must show proper identification for entry.

b. When passengers arrive at the facility, they will disembark and enter the building through the side entrance. Once inside, a security area will be set up where you will show the required identification. You will then be issued a facility badge and will proceed to your designated work area.

4. Clothing and Personal Items:

All personnel are encouraged to bring, as a minimum, two complete changes of clothing, two pairs of shoes, personal hygiene items, a flashlight with extra batteries, a minimum 2-week supply of prescription medications and toiletries, and two sets of prescription eyeglasses.

a. Civilian Personnel: Civilian casual and PT clothing are recommended for off-duty wear. Consideration for weather conditions and seasons should be taken into account for proper outerwear.

b. On Duty, Civilian Personnel: Business casual dress clothing will be worn. Examples: collared shirt and dress slacks for men; blouse or sweater and dress slacks or skirt for women; and appropriate shoes. On-duty dress does not include T-shirts, casual shorts, halter-tops, or flip-flops. Jeans may be worn upon supervisor approval.

c. Off Duty, all Personnel: Off duty personnel are authorized to wear casual civilian clothing provided they are in good taste and repair.

5. Security Procedures

Visiting the Site is unlike any other official visit. All personnel visiting this Site must be listed on their organization's current roster. A copy of each organization's roster should be on file with the facility Security Manager.

Rosters must be updated quarterly to ensure arriving personnel are efficiently in-processed. To facilitate last minute substitutions, each organization will designate an individual who can verify security clearances for personnel not listed on rosters. The Site's Security Manager is the primary point of contact for changes to access rosters.

The Alternate Site is a secure site. All personnel and vehicles entering and exiting the site are subject to inspection and search at all times.

Classified material: To transport any classified material you must have a properly completed XX Form XXXX. All classified material must be properly secured in an approved container whenever the office is unattended and be accountable at all times. All classified material brought into or removed from this facility will be coordinated through the security manager, Ms. Jane Doe, XXX-XXX-XXXX.

a. Badges. Designated personnel visiting this facility will already have a Company badge. For those that do not have a facility badge, they will be issued a badge at the security area prior to entry into the facility. The badge must be worn at all times while in the building. Badges are to be worn clipped to the left side of the collar or left breast pocket of your outer garment. The single exception is that badges will be worn face down by personnel presenting briefings that are being videotaped or while being photographed.

b. After Hours, Weekends, and Holiday Access. Coordination with the Duty person or facilities personnel will be required; a list of contact numbers is on page 12.

c. Prohibited Items. The following items are prohibited and will not be allowed inside the building: Weapons of any kind, firearms, knives (with blades exceeding 3 inches), martial arts weapons, alcoholic beverages, unauthorized drugs, cameras, and transmitting/receiving or recording devices of any kind (i.e., radios, CD recorders, tape recorders, personal laptops and televisions). The Security/Facilities personnel will confiscate prohibited items. Security or facilities personnel are not responsible for confiscated items.

d. Classified Material Destruction. All classified waste generated will be placed in a classified material waste bag marked with red stripes. Site personnel will properly dispose of this material. The facility maintains several approved shredders for FOUO and up to and including secret material.

f. Temporary Classified Documents Storage. All classified documents will be stored in GSA approved containers. This container is located in the secure room in Bldg. XXXX. Proper storage will be provided for all classified documents. All classified material must be accounted for at all times and is the responsibility of the individual creating or holding the documents.

g. Securing Valuables. There is no designated secure storage for personal items.

h. THREAT CONDITIONS:
Entry onto the Site Compound during THREAT CONDITIONS is authorized for essential personnel only. All designated personnel with a facility badge with an EMG sticker will be considered essential.

When THREAT CONDITIONS are in effect, all personnel without a facility badge will have to gather at the pass office by the main gate and contact a facility person to escort them onto the site. To contact a facility person call the emergency/notification number for the EOC listed on page XX.

NOTE: PLANNING ASSUMPTIONS, VULNERABILITIES, OR CONTINUITY PLAN DETAILS ARE SECRET AND SHOULD ONLY BE DISCUSSED WITH THOSE WHO HAVE THE PROPER CLEARANCE LEVEL AND "NEED TO KNOW."

NOTE: IF YOU ARE APPROACHED OR CONTACTED BY REPORTERS, REFER THEM TO THE PUBLIC AFFAIRS OFFICE.

"OPSEC STATEMENT"

To protect the operations and activities of this Facility, you are reminded when you leave this facility to avoid conversations regarding your duties and or the activities and operations. The following Essential Elements of Friendly Information (EEFI) have been identified as sensitive aspects of our facility and require protection from unauthorized disclosure. EEFI should not be discussed outside the facility or disposed of as unclassified trash. The responsibility to protect this information rests with Site personnel of this facility, at all levels.

1. NUMBER/TYPE OF SECURITY PERSONNEL, AND EQUIPMENT
2. OPERATIONAL INFORMATION
3. INFORMATION SYSTEMS INFORMATION
4. SIGNIFICANT VISITORS BY NAME OR TITLE
5. CONSTRUCTION PROJECTS AND MAINTENANCE INFORMATION
6. THE EXISTENCE, LOCATION, AND BASIC MISSION OF THE SITE IS \\ UNCLASSIFIED //
7. FOUO SHOULD BE PROTECTED AND DISCUSSED ONLY FOR OFFICIAL REASONS
8. SPECIFIC CAPABILITIES, CAPACITIES, DESIGN DETAILS, OR COMPLETE LIST OF
9. RELOCATING AGENCIES IS - CONFIDENTIAL AND SHOULD ONLY BE DISCUSSED WITH THOSE WHO HAVE THE PROPER CLEARANCE LEVEL AND "NEED TO KNOW"

FORCE PROTECTION CONDITIONS (FPCON)

FPCON Definitions:

NORMAL: General Global Threat of possible terrorist activity.

ALPHA: General warning, nature and extent unpredictable.

BRAVO: Increased and more predictable threat of terrorist activity exists, no particular target identified.

CHARLIE: Incident occurred or intelligence received that terrorist action is imminent.

DELTA: In the immediate area where an attack has occurred or intelligence received that terrorist action against a specific location is imminent.

5. Facility Emergency Procedures

The following procedures and general information is to be used as a guide to safely respond to emergency situations.

a. Key Personnel and Emergency Telephone List

Designated Building Official- Mr. _____ Phone # (XXX-XXX-XXXX)
Alternate Designatee – Ms. _____ Phone # (XXX-XXX-XXXX)
Occupancy Emergency Coordinator - Ms. _____ Phone # (XXX-XXX-XXXX)
Alternate Emergency Coordinator – Mr. _____ Phone # (XXX-XXX-XXXX)
Security Manager – Ms. _____ Phone # (XXX-XXX-XXXX)
LAN Operations – Mr. _____ Phone # (XXX-XXX-XXXX)
Facilities Manager – Mr. _____ Phone # (XXX-XXX-XXXX)
Sites COOP Coordinator – Mrs. _____ Cell # (XXX-XXX-XXXX)
Telephones, install/repair – Mr. _____ Phone # (XXX-XXX-XXXX)
Facility Emergency notification number – Hotline # (XXX-XXX-XXXX)

FOR EMERGENCY ENTRY-AFTER HOURS AND HOLIDAYS CONTACT: Hotline - XXX-XXX-XXXX or Shift Supervisor at XXX-XXX-XXXX.

b. Immediate Actions for Fire Emergencies

1. Remain Calm / Stay Alert.

2. Activate fire alarm, notify the Fire Dept. (911) and notify Security XXXX.

3. Secure your work areas following security procedures for securing classified items & material.

4. Close all doors; <u>do not lock them.</u> Evacuate the building through designated Emergency Exits. <u>Do Not Use the Elevators</u> and report to your appropriate assembly area for roll call.

5. Duty section will account/assist all impaired (YOUR ORGANIZATION) persons to assembly areas.

6. Remain in your assembly area until otherwise directed by proper authorities.

<u>Do Not Activate the Fire Alarm in the Event of a Bomb Threat!</u>

c. Bomb Threat Procedures

Bomb threats may be received in several ways, by telephone, by mail, or by messenger.

If you receive a Bomb Threat by telephone:

1. Keep the caller on the line as long as possible; talk to the caller; ask him/her to repeat the message.

2. Try to find out:

WHEN…………………………	when will it go off?
WHERE…………………………	where is it, where do we look?
WHAT…………………………	what does it look like?
WHY…………………………	why are you doing this?
WHO…………………………	who are you; what group do you belong to?
WHERE…………………………	where are you?

3. Record the time and date of the call.

4. Pay particular attention to background noises such as motors running, music, laughter, etc.

5. Listen closely to the voice (male, female), voice quality (calm, excited), accents, and speech impediments.

6. Report the call immediately to the Security Manager, XXXX.

d. Chemical, Biological, Radiological, Nuclear Events (CBRN)

1. Move away from all windows, doors, openings to the outside, and ventilation supply outlets.

2. Inform fellow Employees and Supervisory Management of the conditions as known to you at the time.

3. Ensure Building Management (through your Supervisor) is aware of and is taking action to shut down all air handlers and Building air intakes immediately.

4. Evacuate or Shelter in place as directed by Key Personnel. Await further instructions as provided by emergency notification networks.

NOTE: The above is provided as guidance only. The actual event may require other emergency measures.

e. Tornado or Other Severe Weather Conditions

1. Move away from windows and doors. Move toward the center core of the building.

2. Remain calm.

3. Monitor local weather forecasts.

4. Status of Federal Government Operations during severe weather conditions will be broadcast on local broadcast channels.

5. Follow directions of the Key Personnel and Senior Officials.
6. The Web Site will post operating status of the Company or call the <u>Hot Line</u> at XXX-XXX-XXXX.
7. The Site Personnel will post operating status of the Site, XXX-XXX-XXXX or toll free XXX-XXX-XXXX.

NOTE: The Designated Building Official will inform personnel of any early work release.

f. Utility Failure

To report Power failures, call XXXX.

1. Await further guidance from Key Personnel.

2. Should a Utility failure occur during non-working hours, Access to the Building will be denied to all persons.

g. Sheltering In Place <u>Immediate or Imminent Threat</u>

1. Employees must immediately move away from all doors, windows, and ventilation systems that circulate outside air into the building.

Used when a dangerous threat is imminent that could result in sickness, injury, or loss of life.

Emergency First Responders will provide direction compound upon the nature of the emergency to ensure all steps are taken to avoid loss of life or serious injury by way of airborne contaminates, blast, or heat effects of an unknown, known or undetermined threat or threat source.

h. Sheltering In Place <u>Calm/Non-Life Threatening</u>

This sheltering category reduces the amount of vehicular traffic on area roads. This situation is normally used to allow local emergency vehicles to rush equipment and personnel to the site where an incident has occurred.

The situation causing the sheltering in place order has not been deemed as dangerous as to be harmful if employees remain in place.

Employees are free to move about <u>inside</u> the building to conduct normal business.

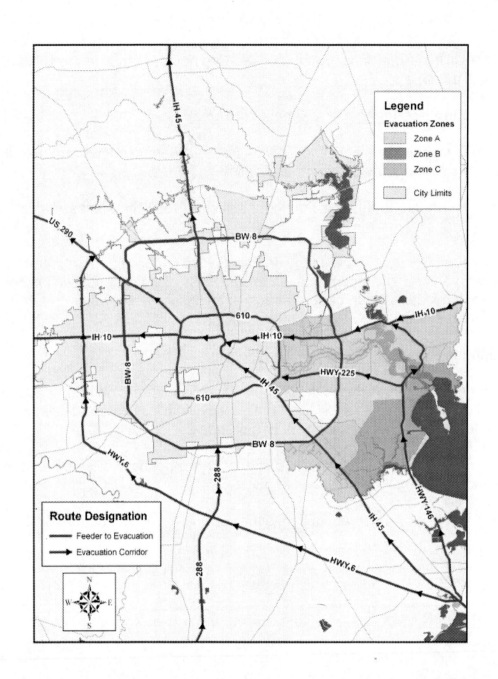

SUPPORT ACTIVITY
EMERGENCY EVACUATION PLAN AND PROCEDURES

Should an event occur requiring the evacuation of the compound or individual buildings, the following guidance is provided. In order for a safe and orderly evacuation, it is imperative to follow the procedures below. A color-coded detailed map is provided as the routes for the following evacuations:

Section I. Compound Evacuation

Section II. Building Evacuations

Section III. Gate Closures

Section IV. Evacuation Routes

Section V. Individuals with disabilities

Employees should become familiar with the location of all stairways, exits, and the nearest evacuation route (posted throughout the building) and become familiar with the location of the nearest building alarm manual pull stations and their operation.

SECTION I: COMPOUND EVACUATION (100% or Partial)

Expect to be advised by one of the following methods:
Building public address system.
Police vehicles announcing evacuation order.
Chain of Command.
Close windows and doors behind you, turning off all electronic equipment and lights as you leave, time permitting.
Take only those articles necessary to evacuate the compound. Exiting the compound will depend on where you park. For example, if you find your parking spot in the red zone, you will exit by the red zone or Gate Main. Please do not cross into and exit another color zone. Police support at the East and South Gates prior to the evacuation order has been coordinated.
If you carpool and miss your ride or use mass transit, report to Building 111 Auditorium for further assistance.

SECTION II: BUILDING EVACUATIONS (Single or Multiple)

Site policies and procedures require that all persons, when directed either by fire alarm or by public address, evacuate their facility immediately. Persons with disabilities or those who require special assistance may not be able to evacuate unassisted. If you require assistance, please follow the directions in Section V.

The need to evacuate a building(s) will be issued the same way a compound evacuation would be ordered. At this time or just outside the building, personnel will be given more information concerning the emergency.

The orange-colored areas identified on the map as A, B, C, D, E, and F will be used to temporarily relocate/evacuate personnel until the emergency has subsided. Personnel will be advised which area to report to at the time of the emergency.

Personnel may also be relocated to the Forum at Building 06 or the Auditorium at Building 201 should conditions permit.

The Incident Assistance Team and Security personnel will assist in the evacuation, tend to personnel needs, and organize personnel accountability efforts.

SECTION III: GATE CLOSURES

Please enter the traffic pattern at a point closest to your parking space and exit at the first open gate. In the event of a closed gate, you will be advised of the closure and directed to continue following the bold black line on the map.

You may have to pass the closed gate(s). Do not stop or attempt to exit. This will only cause more traffic issues.

If Security sees traffic backing up, you may be directed onto Perimeter Road for the next available exit from the compound.

SECTION IV: EVACUATION ROUTES

The installation has four gates. Below are the main connectors or roadways accessible from each gate:

Main Gate
East Gate
South Gate
West Gate

SECTION V: INDIVIDUALS WITH YOUR ORGANIZATIONBILITIES

Individuals who need assistance may voluntary self-identify by completing the attached Form and forwarding it to the servicing Human Resources Office Deputy Equal Employment Opportunity Officer and/or People with disabilities Program Manager. The information is voluntary and will be kept strictly confidential and shared only with those with emergency response duties.

An individual does not have to identify as a person with a permanent disability to qualify for egress assistance. Individuals with reduced stamina, fatigue, emotional, cognitive, thinking or learning difficulties, use of technology or medications, episodic or temporary limitations (stroke recovery, asthma, heart disease, surgery, accidents, and pregnancy) may require evacuation assistance.

Management is responsible for employees' safety and should ensure that the employee receives necessary assistance once evacuated from the building. Senior management will ensure that all individuals are evacuated from the building. Managers of employees with special needs should ensure that the employee is aware of all emergency procedures.

Hearing

Most buildings are equipped with fire alarm, horn/bells, and strobes that sound when a fire detection or manual pull station is pulled. The strobe lights are for hearing-impaired persons. Persons with hearing impairments may not notice or hear emergency alarms and will need to be alerted of emergency situations. Hearing impaired or not you should always be alert and aware of your surroundings. When it appears your co-workers are leaving the building, you should assume there might be an issue and you should follow. Watch for co-workers who don't seem concerned and inform them of the evacuation order.

Visual

Most buildings are equipped with fire alarm, horn/bells, and strobes that sound when a fire detection or manual pull station is pulled. The horn/bell is for sight-impaired persons. Most people with a visual impairment will be familiar with their immediate surroundings and frequently traveled routes. Since the emergency evacuation route is likely different from the commonly traveled route, persons who are visually impaired may require assistance. You should report to your assembly location outside your building where further assistance will be provided.

Mobility Impaired

Persons with mobility impairment(s) must also evacuate during an emergency. Knowing your handicap exit location or the location of an elevator (these are specifically for use during an emergency) will enhance your ability to successfully evacuate during an emergency. If you are unable to evacuate, you should call either 911 or X0606 and identify yourself and building/bay location as soon as possible. This information will be relayed to our on scene responders so that they may enter and assist you.

In order for this plan to work, you will need to practice it. You should know alternate routes home and also how you will communicate with your family in the unlikely event you couldn't get home. On occasion, try going out a different gate and taking a new route home. By doing so, our plan will work.
SUPPORT ACTIVITY

Name:

Telephone Number:

Building and Office Number:

Please describe the type of special assistance needed during an emergency evacuation of the building and/or compound.

6. Communications

Quick Reference Phone Numbers Commercial: Dial XX, number
Facility
Commercial: to call into the facility or compound.
On site, dial last 4 - XXXX
DSN 88-XXX-XXXX
Email: last name & first initial @Alter.Site.Net

Hobby Shop ....
Barber Shop ....
Compound Status Line ....
Clerical Issues ....
Dining ....
Fire and Medical Services
Emergency 911
Fire Dept (Non-Emergency ....
Fitness Center ....
Food Court ....
Gates
East ....
Main ....
South ....
West ....
Golf Course ....
Mail ....
Medical ....
Physical Security ....
Police (Emergency 911

7. Family Readiness

The reasons for emergency deployment are varied ranging from natural disaster to strategic attack. You should prepare your family for all situations that may require them to evacuate while you are away.

a. Develop several emergency plans that can easily be followed without your presence. An evacuation plan and a shelter in place plan for starters.

b. Make note of emergency centers such as local schools, military installations, churches, and hospitals.

c. An excellent reference for Disaster preparedness is available at redcross.org or FEMA.gov.

d. Ensure that your loved ones have a means to communicate with other family members and friends. Family members should know how to call and send mail to you at the site in the event of extreme family emergency.

e. When a disaster strikes, your family might not have access to food, water, and electricity for days, or even weeks. By taking some time now to store emergency food and water supplies, you can provide for your entire family.

f. Talk with your family and discuss why they need to prepare for disaster. Discuss the types of disasters that are most likely to happen. Explain what to do in each situation. Pick two places for your family to meet:

(1) Right outside your home in case of a sudden emergency, like a fire.
(2) Outside your neighborhood, in case you can't return home. Everyone must know his or her address and phone number. Ask an out-of-state friend to be your "family contact."

g. After a disaster, it's often easier to call long distance as most or all <u>LOCAL</u> telephone and cell phone lines will be overloaded. Other family members should call this person and tell them where they are. Everyone must know your contact's phone number. Discuss what to do in an evacuation. Plan how to take care of your pets.

8. Medical and Dental Care

Medical Services
For Emergency Medical care, the Site Fire and Emergency Services provide day-to-day emergency medical treatment and transportation of injured to the local hospital. For emergencies, call 911 from any facility phone. The Site medical clinic is not equipped for emergencies.
The Site dispensary is located in building 777 and provides routine services for military personnel and is operated only during normal working hours.
All personnel should bring an adequate supply of personal prescription medication with them. Recommended minimum 2-week supply along with an extra pair of glasses.
If you have any special medical needs, like special medications, etc., then discuss this matter with your supervisor before visiting the facility. This will allow planning for and meeting those needs in a timely manner.

Dental Services
Emergency Dental services are only available from civilian providers off compound.

The Dental Clinic at _____ provides routine services to personnel and only during normal working hours. They are not equipped for emergencies.

XXXX Hospital
0606 S. Front Street
Pamala, MD 20991
555-XXX-XXXX

From Route XXX or South
Take exit 1906,
Continue past the hospital and immediately after the railroad underpass, left onto Mary Street
Parking garage will be on your left

From Route I-95 North or South
Take exit 32, South Park Street
Travel south on Front Street, approximately five miles
Proceed through five traffic lights and past the Hospital
Immediately after the railroad underpass, left onto Mary Street
Parking garage will be on your left

9. Lodging

Lodging is only available off-site in the local area as follows:

Holiday Inn	Toll Free: 800-5555-Home Fax: XXX-555-5555 1906 Alpha Way Norway, MD 20992
Comfort Inn West	
Hampton Inn	
Hilton	
Days Inn	
Radisson	

East/North – Route 06

1 – T J Maxx
　　Dunham's
　　Outback Restaurant
　　Game Stop
1 – Mexican Restaurant
3 – Rite Aid Pharmacy

4 – CVS Pharmacy
3 – North Philly Steak and Hoagie
6 – Damon's Grille
7 – AMC Center 19 Theatres
8 – Kenny's Center
Kendrid's Food
Essence Crafts
Dollar General Store
Subway
9 – Ruby Tuesday
10 – Red Lobster

East/North – Route 06

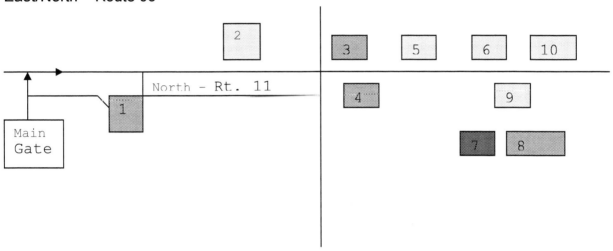

West/South – Rt. 06

1 – Main Gate
2 – Country Buffet Restaurant
3 – Burger King
4 – Bob Evans Restaurant
5 – Dunkin Donuts
6 – Evergreen Chinese Restaurant
7 – Holiday Inn/Legends Restaurant
8 – Hooters
9 – Applebee's
10 – Comfort Inn
11 – Shopping Plaza
    McDonalds
    K-Mart
    Circuit City
    Pet Smart

        Dick's Sporting Goods
        Staples
        Home Depot
12 – Kohl's Department Store
13 – Bowling Alley
14 – Spring Commons
        Olive Garden
Taco Bell
        Florist
        Wal-Mart
        Lowe's
        Dress Barn
        Marshall's
        Pizza Hut
        Silver Spring Diner
15 – Giant Food Market
        Arby's
        Red Robin Restaurant
        Sheetz
        McDonalds
16 – Kentucky Fried Chicken/Long John Silvers

10. Dining Facilities

The site Dining Facilities are located in buildings XXX.

The normal operating hours are:

Breakfast: 0600-1015 (0630 Grill Opens)
Lunch: 1115-1300

There are only two forms of payment accepted in the dining facility: Cash or Meal Card.

Snacks and drinks can be purchased in the break room area in bldg. XXX. Refrigerators are available in the facility break room areas for storage of small amounts of food and drinks.

In the event of a deployment situation or natural disaster emergency, meals will be available. Requests must be submitted in advance for accounting and coordination purposes.

11. Morale Welfare and Recreation

TV and break area are available at the facility.

Barber Shop	Mon/Wed, 0900-1700, Tue/Thu, 0900-1500 Fri, 0900-1400	XXXX
Food Court	Mon-Fri, 0600-1400,	
Fitness Center		
Golf Course	Mon-Fri, 0800-1700,	

12. Departure, Out-Processing

1. All computers will be logged off the network and powered down as applicable. Some Agencies DO NOT power down.
2. All classified material will be accounted for and properly stored.
3. All STU-III, STE keys (if issued) will be accounted for and returned.
4. All security containers will be secured.
5. A security sweep of all spaces will be conducted to ensure all classified material is accounted for.
6. Notify Security and the facility manager of anticipated departure/turnover times to facilitate coordination of your requirements with those of other organizations.
7. At the security post, all personnel will surrender their temporary facility badges.
8. Personnel departing the Site via air will be transported directly to the helipad to meet their aircraft. This should be scheduled as far in advance as possible to allow coordination of air transportation.

Appendix H

Example of Alternate Site Activation,
Reception, and Notification Checklist

The following checklist includes the activities that will need to be completed to ensure activation readiness. These activities should be completed as soon as possible once a COOP notification has been received.

ALERT NOTIFICATION: #1

Dialogic will call "On-Site" Team – Respond to Alert Roster within 30 minutes.
"On-Site" Team will meet at Bldg. 200 within 1 hour of notification, and begin the below checklist, and locate on
If after normal working hours, notify security at (XXXX) that people will be entering Bldg. 200, and some alarmed doors will be accessed.
Retrieve COOP Activation Plan Book from safe in area 201. Take book to EOC the front conference tables for reference by all arriving team members.

INITIAL PROCEDURES:

Completed	Activity
	Contact Company Duty Section personnel and instruct them to report to man the EOC reception desk, and the front security desk.
	Get Enclosure 1 – *"Access Combinations"* from the enclosures section of this binder, along with the set of master keys located in the safe drawer with this binder.
	Turn on lights and unlock relevant areas – 13, 17, 27B, 30, 32, 37, 51, 56, 77, 88A, and 89.
	Get two black plastic chests from room 27B and take one to room 77, and one to room 88A.

RECEPTION: #1

Personnel with on-site badges will enter through the front glass doors by EOC and will need to go through check in. Personnel without site badges should be directed to the Main entrance, security area for check in, and visitor badges.

Completed	Activity
	On-Site Personnel must man EOC glass doors for in-processing procedures until relocation personnel arrive and relieve them.
	Work Station Signs are located in the "EOC" (found in the reception desk).
	Get Personnel Sign-In Sheets (found in EOC reception desk or COOP Activation Plan Book) for in processing.
	In-processing includes: Show ID and sign in using "Personnel Sign In Sheets." Distribute Site Guide (if not already received). Document confirmed hotel name and number
	Direct arriving personnel to locate assigned work area from easel with room layout.

EOC

Completed	Activity
	Get easel from room 33 and display in front area of EOC with seating diagram showing numbering scheme and group layout.
	Get group name table tents from room 27B and distribute to the EOC desktops according to Enclosure 5 – *"EOC Seating Diagram."*
	Remove all microphones from desktops and turn on all phone ringers.
	Get carts of unclassified laptops from room 32, roll into the EOC, and distribute to the EOC desktops according to Enclosure 2 – *"Laptop Location Diagram"*, U (unclassified) 1 through U36. Return carts to room 27B.

Completed	Activity
	Get classified laptops from 5 drawer safe located behind door in room 27B and distribute to the EOC desktops according to Enclosure 2 – *"Laptop Location Diagram"*, C (classified) 1 through C24.
	Get 36 Cat 5 LAN jumper wires from cabinet in room 32, and connect all EOC unclassified laptops to K533 LAN connections on desktops.
	Get 24 fiber jumpers from cabinet in room 32, and connect all EOC classified laptops to fiber LAN connections on desktops.
	Secure Fax – Get one MCI 1300 STUIII from room 007 (MMS) and locate on table in the right front of EOC. Get one fax machine and one secure fax gateway from cabinet in room 27B, and connect all three together and to phone line as shown in Enclosure 3 – *"Secure Fax Setup"* for secure fax capability.
	STUIIIs – will have to be set up "as needed".
	Get smaller copier/printer (JVC AL-3331CS) from behind door in room 32, and set up on table in the right front of EOC. Make LAN connection to copier/printer with KLB-43 jumper connected to printer.
	Get one black and white printer and one color printer from? and locate on table in the right front of EOC. Make LAN connection to printers.

EOC 2:

Completed	Activity
	Get easel from room 51 and display at entrance to 56 with seating diagram of numbering scheme and group layout.
	Get group name displays from room 51 and secure to the room 56 tables according to Enclosure 6 – "*56 Seating Diagram*".
	Get Enclosure 4 – *"Room 56 Setup"* and determine how may tables will be necessary to get as close as possible to the setup diagram taking into consideration how many existing workstations can be utilized. Get needed tables from room 77 and setup as close as possible according to diagram.

Completed	Activity
	Get carts of unclassified laptops from room 77, roll into the 56 and distribute to the 56 desktops according to Enclosure 2 – *"Laptop Location Diagram"*, U (unclassified) 1 through U36. Return carts to room 77.
	Get classified laptops from 5-drawer safe located behind door in room 127A and distribute to the 56 desktops according to Enclosure 2 – *"Laptop Location Diagram"*, C (classified) 1 through C24.
	Get 36 Cat 5 LAN jumper wires from cabinet in room 77, and connect all 56 unclassified laptops to KLB 42 LAN connections on desktops.
	Get 24 fiber jumpers from cabinet in room 77, and connect all 56 classified laptops to fiber LAN connections on desktops.
	Secure Fax – Get one MCI 1300 STUIII from room 32 (PPS) and locate on table designated for "Printers and Fax" in room 17. Get one fax machine and one secure fax gateway from cabinet in room 77, and connect all three together as shown in Enclosure 3 – *"Secure Fax Setup"* for secure fax capability.
	Get two unclassified fax machines from cabinet in room 77 and set up on tables designated for "Printers and Fax" in room 32. Make phone line connections.
	Get two black and white printers and one color printer from ? and locate on tables designated for "Printers and Fax" in room 56. Make LAN connections to printers.
	Get 50" plasma from room 13 and locate on a table in room 17. Make necessary cable connection to display CNN.
	Make necessary cable connection to display CNN on plasma.

EOC PERSONNEL:

Completed	Activity
	Bring up all EOC A/V equipment and display CNN on all display screens.
	Ensure that the 50" plasma in room 17 is connected to the cable connections via VCR and displaying CNN.

Completed	Activity
	Ensure that the plasma in room 17 is connected to the cable connections and displaying CNN.
	Assist with Video Tele Center (VTC) s as necessary.

LAN PERSONNEL:

Completed	Activity
	Make all necessary (classified and unclassified) LAN connections active to 13 and 17.
	Ensure that all classified and unclassified computers are made active and useable in the EOC and room 17.
	Ensure that classified and unclassified computers are made active and useable in room 50 and 56A.
	Ensure that all printers are made active and useable in the EOC and room 17.

PHONE PERSONNEL:

Completed	Activity
	Ensure that all phones are made active and useable in the EOC and room 13.
	Ensure that all fax machines are made active and useable in the EOC and room 13.
	Assist with deployment of STUIIIs in 13 and 17

RECEPTION: #2

The following checklist includes the activities that will need to be completed to ensure in processing is completed.

Completed	Activity
	Personnel will enter through main door and go through in-process which includes the following: Show ID and sign in Receive facility badge Receive lodging assignment and directions Receive dining hours and Facility Guide Personnel asked to call hotel and confirm their room
	Personnel will locate assigned workstation and begin working.

***NOTE: Activation Team personnel will complete in processing.

Alert and Notification Checklist #2

The following checklist includes the activities that will need to be completed when we are notified a COOP has been declared.

Completed	Activity
	Receive telephone call from Front Office and/or Alert Notification System that "Code Big Bird" has begun.
	Activate Alert Notification Process for Organization
	"Site Activation Team meets within 1 hour of notification
	Retrieve Site Activation Plan Book from Safe in Bldg. XXX and return to Bldg. XXX for set-up

Activation Checklist #2

The following checklist includes the activities that will need to be completed to ensure activation readiness. These activities should be completed as soon as possible once a COOP has been activated.

Completed	Activity
	Sensor Lights activated upon entering the building.
	Secure all doors.
	Notify hotels and confirm number of rooms available. Place hotel information on sign-in sheet. Ensure Holiday Inn rooms are assigned to personnel on list marked Holiday Inn guests.
	Obtain safe combination from Appendix A in the Site Activation Plan Book.
	Proceed to west end of building to the warehouse and retrieve unclassified laptop carts.
	Place unclassified laptop at appropriate workstation and plug in.
	Go to safe and retrieve classified laptops.
	Place classified laptops at appropriate workstations.
	Turn TVs on to CNN.
	Put table and four chairs at east entrance door to be used as in-processing area.
	Place easel and seating chart at table.

Appendix I

An example of FEMA, Continuity of Operations (COOP) Plan Template Instructions

Continuity of Operations (COOP) Plan Template Instructions

Federal Emergency Management Agency
500 C ST, SW
Washington, D.C. 20472

GUIDE INSTRUCTIONS

This guide provides instructions for developing a Continuity of Operations (COOP) Plan according to *Department of Homeland Security (DHS) Headquarters Continuity of Operations (COOP) Guidance Document, dated April 2004*. Although general guidance and sample information is provided in this guide for reference, organizations are encouraged to tailor COOP Plan development to meet their own needs and requirements. These instructions accompany an electronic template that may be downloaded from the Federal Emergency Management Agency (FEMA) Office of National Security Coordination (ONSC) website at the following address: http://www.beta.fema.gov/test/ns/index.shtm

Questions concerning this guide can be directed to:

Office of National Security Coordination
Federal Emergency Management Agency
500 C Street, SW
Washington, DC 20472
(202) 646 3754

TABLE OF CONTENTS

I. Executive Summary 143
II. Introduction 143
III. Purpose 143
IV. Applicability and Scope 143
V. Essential functions 143
VI. Authorities and References 144
VII. Concept of Operations 144
A. Level I: Activation and Relocation 144
1. Decision Process 145
2. Alert, Notification, and Implementation Process 145
3. Leadership 145
B. Level II: Alternate Facility Operations 146
1. Mission Critical Systems 146
2. Vital Files, Records, and Datacompounds 147
C. Level III: Reconstitution 149
VIII. COOP Planning Responsibilities 149
IX. Logistics 150
A. Alternate Location 150
B. Interoperable Communications 150
X. Test, Training, and Exercises 150
XI. Multi-Year Strategy and program Management Plan 151
XII. COOP Plan Maintenance 151
Annex A: Authorities and References 151
Annex B: Operational Checklists 151
Annex C: Alternate Location/Facility Information 151

Annex D: Maps and Evacuation Routes 151
Annex E: Definitions and Acronyms 152

Executive Summary
The executive summary should briefly outline the organization and content of the COOP Plan and describe what it is, whom it affects, and the circumstances under which it should be executed. Further, it should discuss the key elements of COOP planning and explain the organization's implementation strategies.

Introduction
The introduction to the COOP Plan should explain the importance of COOP planning to the organization. It may also discuss the background for planning and referencing recent events that have led to the increased emphasis on the importance of a COOP capability for the organization.

Purpose
The purpose section should explain why the organization is developing a COOP Plan. It should briefly discuss applicable Federal guidance and explain the overall purpose of COOP planning, which is to ensure the continuity of mission essential functions. Because of today's changing threat environment, this section should state that the COOP Plan is designed to address the all hazard threat.

Applicability and Scope
This section describes the applicability of the plan to the organization as a whole, headquarters, as well as subordinate activities, co-located and geographically dispersed, and to specific personnel groups of the organization. It should also include the scope of the plan. Ideally, plans should address the full spectrum of potential threats, crises, and emergencies (natural as well as manmade).

Essential functions
The essential functions section should include a list of the organization's prioritized essential functions. Essential functions are those organizational functions and activities that must be continued under any and all circumstances.

Organizations should:
Prioritize these essential functions
Establish staffing and resource requirements
Integrate supporting activities
Develop a plan to perform additional functions as the situation permits.
For additional information on essential functions, see Section X of the DHS HQ COOP Guidance Document.

SAMPLE
The following table shows examples of prioritized essential functions for a fictitious organization, the Bureau of Water Management:

Priority	Essential Functions
1	Administer programs to protect the region's water supply and the health of the public.
2	Ensure the protection of fish and aquatic life.
3	Ensure pollution prevention and compliance assurance.
4	Provide technical support and information to assist in planning and restoration.
5	Approve and oversee cleanups of contaminated sites.
6	Plan and implement regional flood control programs.

Authorities and References
This section should reference an annex that outlines all supporting authorities and references that have assisted in the development of this COOP Plan.
Annex A of the DHS COOP Guidance Document provides a list of authorities and references.

Concept of Operations
This section should explain how the organization will implement its COOP Plan, and specifically, how it plans to address each critical COOP element. This section should be separated into three levels: activation and relocation, alternate facility operations, and reconstitution.

Level I: Activation and Relocation
The Level I section should explain COOP Plan activation procedures and relocation procedures from the primary facility to the alternate facility. This section should also address procedures and guidance for non-relocating personnel.

Decision Process
This section should explain the logical steps associated with implementing a COOP Plan, the circumstances under which a plan may be activated (both with and without warning), and should identify who has the authority to activate the COOP Plan. This process can be described here or depicted in a graphical representation.

Alert, Notification, and Implementation Process
This section should explain the events following a decision to activate the COOP Plan. This includes employee alert and notification procedures and the COOP Plan implementation process.

Leadership
Orders of Succession
This section should identify orders of succession to key positions within the organization. Orders should be of sufficient depth to ensure the organization's ability to manage and direct its essential functions and operations. The conditions under which succession will take place, the method of notification, and any temporal, geographical, or organizational limitations of authority should also be identified in this section.
For additional information on succession, see Section XI.I of the DHS HQ COOP Guidance Document.

SAMPLE
The following table shows the order of succession for the Director of the Bureau of Water Management:

Successors
Director, Bureau of Water Management
Deputy Director, Bureau of Water Management
Division Head, Enforcement and Remediation Division
Division Head, Standards and Planning Division

Delegations of Authority
This section should identify, by position, the authorities for making policy determinations and decisions at headquarters, field levels, and other organizational locations, as appropriate. Generally, pre-determined delegations of authority will take effect when normal channels of direction are disrupted and terminate when these channels have resumed. Such delegations may also be used to address specific competency requirements related to one or more essential functions that are not otherwise satisfied by the order of succession. For example, once again referring to the Bureau of Water Management sample: because of particular technical skill, the authority to remediate contaminated sites could be pre-delegated directly to the Division Head, Enforcement and Remediation Division, bypassing the intervening successor. Delegations of authority should document the legal authority for making key decisions, identify the

programs and administrative authorities needed for effective operations, and establish capabilities to restore authorities upon termination of the event. Pre-determined delegations of authority may be particularly important in a devolution scenario.
For additional information on delegations of authority, see Section XI.2 of the DHS HQ COOP Guidance Document.

Devolution
The devolution section should address how an organization will identify and conduct its essential functions in the aftermath of a worst-case scenario, one in which the leadership is incapacitated. The organization should be prepared to transfer all of their essential functions and responsibilities to personnel at a different office or location.
For additional information on Devolution, see Section XI.3 of the DHS HQ COOP Guidance Document.

Level II: Alternate Facility Operations
The Level II section should identify initial arrival procedures as well as operational procedures for the continuation of essential functions.

Mission Critical Systems
The section should address the organization's mission critical systems necessary to perform essential functions and activities. Organizations must define these systems and address the method of transferring/replicating them at an alternate site.

SAMPLE
The following table shows examples of mission critical systems for the Bureau of Water Management:

System Name	Current Location	Other Locations
Hazardous Spill Cleanup Unit	Warehouse 11	Storage Unit B, Storage Unit C
Mobile Analytical Laboratory	Primary Facility	Alternate Facility
Mobile Operations Center	Primary Facility	Alternate Facility

Vital Files, Records, and Data Compounds
This section should address the organization's vital files, records, and database, to include classified or sensitive data, which are necessary to perform essential functions and activities and to reconstitute normal operations after the emergency ceases. Organizational elements should pre-position and update on a regular basis those duplicate records, database, or back-up electronic media necessary for operations. There are three categories of records to be reviewed and prioritized, then transferred (either hard copy or electronic media) to an alternate location:
Emergency operations records;
Legal/financial records; and,
Records used to perform national security preparedness functions and activities (EO 12656).
For additional information on vital files, records, and data compounds, see Section XV of the DHS HQ COOP Guidance Document.

SAMPLE
The following table shows examples of vital files, records, and database for the Bureau of Water Management:

Vital File, Record, or Database	Form of Record (e.g., hardcopy, electronic)	Pre-positioned at Alternate Facility	Hand Carried to Alternate Facility	Backed up at Third Location
GIS Mapping Database	Electronic	X		X
List of Licensed Spill Cleanup Contractors	Hardcopy		X	
List of Regional Dams	Hardcopy		X	
Pollution/Chemical Incident Database	Electronic	X		
Public and Private Sewage System Records	Electronic	X		X
Water Treatment Regulations	Hardcopy		X	
Flood Control Policies	Hardcopy		X	
Federal Water Facilities Contact List	Hardcopy		X	

The following table shows sample vital files, records, and database pertaining to COOP coordination for the Bureau of Water Management:

Vital File, Record, or Database	Form of Record (e.g., hardcopy, electronic)	Pre-positioned at Alternate Facility	Hand Carried to Alternate Facility	Backed up at Third Location
COOP Plan	Electronic	X		X
Phone Roster	Hardcopy		X	
Devolution Contact List	Hardcopy		X	
Legal Authority List	Electronic	X		
Emergency Water Resources List	Hardcopy		X	

Level III: Reconstitution

The Level III section should explain the procedures for returning to normal operations – a time leveled approach may be most appropriate. This section may include procedures for returning to the primary facility, if available, or procedures for acquiring a new facility through General Services Administration (GSA). Notification procedures for all employees returning to work must also be addressed. The conduct of an After Action Report (AAR), to determine the effectiveness of COOP plans and procedures should be considered.

COOP Planning Responsibilities

This section should include additional delineation of COOP responsibilities of each key staff position, to include individual Emergency Relocation Group (ERG) members, those identified in the order of succession, and delegation of authority, and others, as appropriate.

SAMPLE
The following table shows examples of some COOP responsibilities for the Bureau of Water Management:

Responsibility	Position
Update COOP plan annually.	Division Head, Standards and Planning Division
Update telephone rosters monthly.	Communications Specialist, Standards and Planning Division
Review status of vital files, records, and database.	Records Specialist, Standards and Planning Division
Conduct alert and notification tests.	Communications Specialist, Standards and Planning Division
Develop and lead COOP training.	Training Specialist, Standards and Planning Division
Plan COOP exercises.	Training Specialist, Standards and Planning Division

Logistics
Alternate Location
The alternate location section should explain the significance of identifying an alternate facility, the requirements for determining an alternate facility, and the advantages and disadvantages of each location. Senior managers should take into consideration the operational risk associated with each facility. Performance of a risk assessment is vital in determining which alternate location will best satisfy an organization's requirements. Alternate facilities should provide:
(1) Sufficient space and equipment
(2) Capability to perform essential functions within 12 hours, up to 30 days
(3) Reliable logistical support, services, and infrastructure systems
(4) Consideration for health, safety, and emotional well being of personnel
(5) Interoperable communications
(6) Computer equipment and software
For additional information on alternate facilities, see Section XII of the DHS HQ COOP Guidance Document.

Interoperable Communications
The interoperable communications section should identify available and redundant critical communication systems that are located at the alternate facility. These systems should provide the ability to communicate within the organization and outside the organization.
For additional information on interoperable communications, see Section XIII of the DHS COOP Guidance Document.

Test, Training, and Exercises
This section should address the organization's Test, Training, and Exercise (TT&E) Plan. Tests, Training, and Exercise familiarize staff members with their roles and responsibilities during an emergency, ensure that systems and equipment are maintained in a constant state of readiness, and validate certain aspects of the COOP Plan. Managers may be creative when it comes to COOP readiness and include snow

days, power outages, server crashes, and other ad-hoc opportunities to assess preparedness.
For additional information on TT&E, see Section XVII of the DHS COOP Guidance Document.

Multi-Year Strategy and program Management Plan

This section should discuss how the organization plans to develop their Multi-Year Strategy and Program Management Plan (MYSPMP). The MYSPMP should address short and long term COOP goals, objectives, and timelines, budgetary requirements, planning and preparedness considerations, and planning milestones or tracking systems to monitor accomplishments. It should be developed as a separate document.
For additional information on MYSPMP development, see Section XVIII of the DHS HQ COOP Guidance Document.

COOP Plan Maintenance
This section should address how the organization plans to ensure that the COOP Plan contains the most current information. Federal guidance states that organizations should review the entire COOP Plan at least annually. Key evacuation routes, roster, and telephone information, as well as maps and room/building designations of alternate locations should be updated as changes occur.
For additional information on COOP Plan maintenance, see Section XIX of the DHS HQ COOP Guidance Document.

Annex A: Authorities and References

This annex should cite a list of authorities and references that mandate the development of this COOP Plan, and provide guidance toward acquiring the requisite information contained in this COOP Plan.

Annex B: Operational Checklists

This section should contain operational checklists for use during a COOP event. A checklist is a simple tool that ensures all required tasks are accomplished so that the organization can continue operations at an alternate location. Checklists may be designed to list the responsibilities of a specific position or the steps required to complete a specific task. Telephone Cascade
Sample operational checklists may include:

Annex C: Alternate Location/Facility Information

This annex should include general information about the alternate location/facility. Examples include the address, points of contact, and available resources at the alternate location.

Annex D: Maps and Evacuation Routes

This annex should provide maps, driving directions, and available modes of transportation from the primary facility to the alternate location. Evacuation routes from the primary facility should also be included.

Annex E: Definitions and Acronyms
This annex should contain a list of key words, phrases, and acronyms used throughout the COOP Plan and within the COOP community. Each key word, phrase and acronym should be clearly defined.

Appendix J

MEF Analysis

MEF #	MEF Descrip	Agency	Priority			RTO In Hours					Req Personnel	Req Systems	Req Software	Alternate Sites		
			1 High	2 Med	3 Low	0-4	4-8	12	24	24+				1	2	3

Appendix K

Business Impact Analysis (BIA)

Corporate Focus

- ✓ Saving Lives "Employee, Customers and Etc…
- ✓ Saving Property "Asset Protection"
- ✓ The Company's Image
- ✓ Employee Morale
- ✓ The Competitive Edge

BIA Decision Metric

Your Organization ↓	Company Focus					
	Saving Lives	Saving Property	Company's Image	Employee Morale	Competitive Edge	Over All
Senior Management	■	▨	■	▨	■	▨
Middle Management	■	■	■	■	▨	■
Front Line Management	■	▨	▨	■	■	■
Employees	■	■	■	■	■	■
Employees	■	▨	▨	▨	■	▨
Over All	■	▨	▨	▨	■	■
	Priority 1	Priority 5	Priority 4	Priority 2	Priority 3	

Priority ➝ HIGH | MEDIUM | LOW

155

"The Concept of Operation"
Is
Based on Tangible and Intangible Losses.

Senior Management's & Over-all BIA Priorities Metric must
Establishes Your Organizations priorities and focus.

- Tangible
    - People
    - Property
    - Inventory

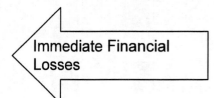

- Intangible
    - Employee Morale
    - The Company's Image
    - Competitive Edge

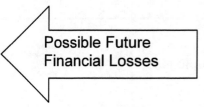